W9-CJS-393

ARK PAPERBACKS

PURITY AND DANGER

This remarkable book, which is written in a very graceful, lucid and polemical style, is a symbolic interpretation of the rules of purity and pollution. Mary Douglas shows that to examine what is considered as unclean in any culture is to take a looking-glass approach to the ordered patterning which that culture strives to establish. Such an approach affords a universal understanding of the rules of purity which applies equally to secular and religious life and equally to primitive and modern societies.

MARY DOUGLAS

Mary Douglas is a distinguished international anthropologist who is currently Professor of Anthropology at Northwestern University, Illinois.

ARK

MARY DOUGLAS

PURITY AND DANGER

AN ANALYSIS OF THE CONCEPTS OF POLLUTION AND TABOO

ARK PAPERBACKS

London and New York

First published in 1966
ARK Edition 1984
Reprinted in 1985 and 1988
ARK PAPERBACKS is an imprint of
Routledge & Kegan Paul Ltd

11 New Fetter Lane, London EC4P 4EE

Published in the USA by
Routledge & Kegan Paul Inc.,
in association with Methuen Inc.,
29 West 35th St., New York, NY 10001

Printed and bound in Great Britain by
Cox & Wyman Ltd, Reading

ISBN 0-7448-0011-0

CONTENTS

ACKNOWLEDGMENTS

I WAS FIRST interested in pollution behaviour by Professor Srinivas and the late Franz Steiner who each, as Brahmin and Jew, tried in their daily lives to handle problems of ritual cleanness. I am grateful to them for making me sensitive to gestures of separation, classifying and cleansing. I next found myself doing fieldwork in a highly pollution-conscious culture in the Congo and discovered in myself a prejudice against piecemeal explanations. I count as piecemeal any explanations of ritual pollution which are limited to one kind of dirt or to one kind of context. My biggest debt for acknowledgement is to the source of this prejudice which forced me to look for a systemic approach. No particular set of classifying symbols can be understood in isolation, but there can be hope of making sense of them in relation to the total structure of classifications in the culture in question.

The structural approach has been widely disseminated since the early decades of this century, particularly through the influence of Gestalt psychology. It only made its direct impact on me through Professor Evans-Pritchard's analysis of the political system of the Nuer (1940).

The place of this book in anthropology is like the invention of the frameless chassis in the history of car-design. When the chassis and the body of the car were designed separately the two were held together on a central steel frame. In the same way political theory used to take the organs of central government as the frame of social analysis: social and political institutions could be considered separately. Anthropologists were content to describe primitive political systems by a list of official titles and assemblies. If central government did not exist, political analysis was held

Acknowledgments

irrelevant. In the 1930's car designers found that they could eliminate the steel frame if they treated the whole car as a single unit. The stresses and strains formerly carried by the frame are now able to be carried by the body of the car itself. At about the same time Evans-Pritchard found that he could make a political analysis of a system in which there were no central organs of government and in which the weight of authority and the strains of political functioning were dispersed through the whole structure of the body politic. So the structural approach was in the air of anthropology before Levi-Strauss was stimulated by structural linguistics to apply it to kinship and mythology. It follows that anyone approaching rituals of pollution nowadays would seek to treat a people's ideas of purity as part of a larger whole.

My other source of inspiration has been my husband. In matters of cleanness his threshold of tolerance is so much lower than my own that he more than anyone else has forced me into taking a stand on the relativity of dirt.

Many people have discussed chapters with me and I am very grateful for their criticism, particularly the Bellarmine Society of Heythrop College, Robin Horton, Father Louis de Sousberghe, Dr. Shifra Strizower, Dr. Cecily de Monchaux, Professor Vic Turner and Dr. David Pole. Some have been kind enough to read drafts of particular chapters and comment on them: Dr. G. A. Wells on Chapter 1, Professor Maurice Freedman on Chapter 4, Dr. Edmund Leach, Dr. Ioan Lewis and Professor Ernest Gellner on Chapter 6, Dr. Mervyn Meggitt and Dr. James Woodburn on Chapter 9. I am particularly grateful to Professor S. Stein, Head of the Department of Hebrew Studies in University College for his patient corrections of an early draft of Chapter 3. He has not seen the final version and is not responsible for further mistakes in biblical scholarship which may have crept in. Nor is Professor Daryll Forde who has frequently read early versions of the book responsible for the final result. I am specially grateful for his criticisms.

This book represents a personal view, controversial and often premature. I hope that the specialists into whose province the argument has flowed will forgive the trespass, because this is one of the subjects which has hitherto suffered from being handled too narrowly within a single discipline.

M.D.

INTRODUCTION

THE NINETEENTH CENTURY saw in primitive religions two peculiarities which separated them as a block from the great religions of the world. One was that they were inspired by fear, the other that they were inextricably confused with defilement and hygiene. Almost any missionary's or traveller's account of a primitive religion talks about the fear, terror or dread in which its adherents live. The source is traced to beliefs in horrible disasters which overtake those who inadvertently cross some forbidden line or develop some impure condition. And as fear inhibits reason it can be held accountable for other peculiarities in primitive thought, notably the idea of defilement. As Ricoeur sums it up:

> 'La souillure elle-même est à peine une
> representation et celle-ci est noyée dans une
> peur spécifique qui bouche la réflexion; avec
> la souillure nous entrons au règne de la Terreur.'
>
> (p. 31)

But anthropologists who have ventured further into these primitive cultures find little trace of fear. Evans-Pritchard's study of witchcraft was made among the people who struck him as the most happy and carefree of the Sudan, the Azande. The feelings of an Azande man, on finding that he has been bewitched, are not terror, but hearty indignation as one of us might feel on finding himself the victim of embezzlement.

The Nuer, a deeply religious people, as the same authority points out, regard their God as a familiar friend. Audrey Richards, witnessing the girls' initiation rites of the Bemba, noted the casual, relaxed attitude of the performers. And so the tale goes on. The anthropologist sets out expecting to see rituals performed with reverence, so say the least. He finds himself in the role of the agnostic sightseer in St. Peter's, shocked at the disrespectful clatter of the adults and the children playing Roman shovehalfpenny on the floor stones. So primitive religious fear, together with the idea that it blocks the functioning of the mind, seems to be a false trail for understanding these religions.

Hygiene, by contrast, turns out to be an excellent route, so long as we can follow it with some self-knowledge. As we know it, dirt is essentially disorder. There is no such thing as absolute dirt: it exists in the eye of the beholder. If we shun dirt, it is not because of craven fear, still less dread or holy terror. Nor do our ideas about disease account for the range of our behaviour in cleaning or avoiding dirt. Dirt offends against order. Eliminating it is not a negative movement, but a positive effort to organise the environment.

I am personally rather tolerant of disorder. But I always remember how unrelaxed I felt in a particular bathroom which was kept spotlessly clean in so far as the removal of grime and grease was concerned. It had been installed in an old house in a space created by the simple expedient of setting a door at each end of a corridor between two staircases. The decor remained unchanged: the engraved portrait of Vinogradoff, the books, the gardening tools, the row of gumboots. It all made good sense as the scene of a back corridor, but as a bathroom— the impression destroyed repose. I, who rarely feel the need to impose an idea on external reality, at least began to understand the activities of more sensitive friends. In chasing dirt, in papering, decorating, tidying we are not governed by anxiety to escape disease, but are positively re-ordering our environment, making it conform to an idea. There is nothing fearful or un-reasoning in our dirt-avoidance: it is a creative movement, an attempt to relate form to function, to make unity of experience. If this is so with our separating, tidying and purifying, we should interpret primitive purification and prophylaxis in the same light.

In this book I have tried to show that rituals of purity and impurity create unity in experience. So far from being aberrations from the central project of religion, they are positive contributions to atonement. By their means, symbolic patterns are worked out and publicly displayed. Within these patterns disparate elements are related and disparate experience is given meaning.

Pollution ideas work in the life of society at two levels, one largely instrumental, one expressive. At the first level, the more obvious one, we find people trying to influence one another's behaviour. Beliefs reinforce social pressures: all the powers of the universe are called in to guarantee an old man's dying wish, a mother's dignity, the rights of the weak and innocent. Political power is usually held precariously and primitive rulers are no exception. So we find their legitimate pretensions backed by beliefs in extraordinary powers emanating from their persons, from the insignia of their office or from words they can utter. Similarly the ideal order of society is guarded by dangers which threaten transgressors. These danger-beliefs are as much threats which one man uses to coerce another as dangers which he himself fears to incur by his own lapses from righteousness. They are a strong language of mutual exhortation. At this level the laws of nature are dragged in to sanction the moral code: this kind of disease is caused by adultery, that by incest; this meteorological disaster is the effect of political disloyalty, that the effect of impiety. The whole universe is harnessed to men's attempts to force one another into good citizenship. Thus we find that certain moral values are upheld and certain social rules defined by beliefs in dangerous contagion, as when the glance or touch of an adulterer is held to bring illness to his neighbours or his children.

It is not difficult to see how pollution beliefs can be used in a dialogue of claims and counter-claims to status. But as we examine pollution beliefs we find that the kind of contacts which are thought dangerous also carry a symbolic load. This is a more interesting level at which pollution ideas relate to social life. I believe that some pollutions are used as analogies for expressing a general view of the social order. For example, there are beliefs that each sex is a danger to the other through contact with sexual fluids. According to other beliefs only one sex is endangered by contact with the other, usually males from females, but sometimes the reverse. Such patterns of sexual danger can be seen to express symmetry or hierarchy. It is implausible to interpret them as expressing something about the actual relation of the sexes. I suggest that many ideas about sexual dangers are

3

better interpreted as symbols of the relation between parts of society, as mirroring designs of hierarchy or symmetry which apply in the larger social system. What goes for sex pollution also goes for bodily pollution. The two sexes can serve as a model for the collaboration and distinctiveness of social units. So also can the processes of ingestion portray political absorption. Sometimes bodily orifices seem to represent points of entry or exit to social units, or bodily perfection can symbolise an ideal theocracy.

Each primitive culture is a universe to itself. Following Franz Steiner's advice in *Taboo*, I start interpreting rules of uncleanness by placing them in the full context of the range of dangers possible in any given universe. Everything that can happen to a man in the way of disaster should be catalogued according to the active principles involved in the universe of his particular culture. Sometimes words trigger off cataclysms, sometimes acts, sometimes physical conditions. Some dangers are great and others small. We cannot start to compare primitive religions until we know the range of powers and dangers they recognise. Primitive society is an energised structure in the centre of its universe. Powers shoot out from its strong points, powers to prosper and dangerous powers to retaliate against attack. But the society does not exist in a neutral, uncharged vacuum. It is subject to external pressures; that which is not with it, part of it and subject to its laws, is potentially against it. In describing these pressures on boundaries and margins I admit to having made society sound more systematic than it really is. But just such an expressive over-systematising is necessary for interpreting the beliefs in question. For I believe that ideas about separating, purifying, demarcating and punishing transgressions have as their main function to impose system on an inherently untidy experience. It is only by exaggerating the difference between within and without, above and below, male and female, with and against, that a semblance of order is created. In this sense I am not afraid of the charge of having made the social structure seem over-rigid.

But in another sense I do not wish to suggest that the primitive cultures in which these ideas of contagion flourish are rigid, hide-bound and stagnant. No one knows how old are the ideas of purity and impurity in any non-literate culture: to members they must seem timeless and unchanging. But there is every reason to believe that they are sensitive to change. The same

4

impulse to impose order which brings them into existence can be supposed to be continually modifying or enriching them. This is a very important point. For when I argue that the reaction to dirt is continuous with other reactions to ambiguity or anomaly, I am not reviving the nineteenth century hypothesis of fear in another guise. Ideas about contagion can certainly be traced to reaction to anomaly. But they are more than the disquiet of a laboratory rat who suddenly finds one of his familiar exits from the maze is blocked. And they are more than the discomfiture of the aquarium stickleback faced with an anomalous member of his species. The initial recognition of anomaly leads to anxiety and from there to suppression or avoidance; so far, so good. But we must look for a more energetic organising principle to do justice to the elaborate cosmologies which pollution symbols reveal.

The native of any culture naturally thinks of himself as receiving passively his ideas of power and danger in the universe, discounting any minor modifications he himself may have contributed. In the same way we think of ourselves as passively receiving our native language and discount our responsibility for shifts it undergoes in our life time. The anthropologist falls into the same trap if he thinks of a culture he is studying as a long established pattern of values. In this sense I emphatically deny that a proliferation of ideas about purity and contagion implies a rigid mental outlook or rigid social institutions. The contrary may be true.

It may seem that in a culture which is richly organised by ideas of contagion and purification the individual is in the grip of iron-hard categories of thought which are heavily safeguarded by rules of avoidance and by punishments. It may seem impossible for such a person to shake his own thought free of the protected habit-grooves of his culture. How can he turn round upon his own thought-process and contemplate its limitations? And yet if he cannot do this, how can his religion be compared with the great religions of the world?

The more we know about primitive religions the more clearly it appears that in their symbolic structures there is scope for meditation on the great mysteries of religion and philosophy. Reflection on dirt involves reflection on the relation of order to disorder, being to non-being, form to formlessness, life to death. Wherever ideas of dirt are highly structured their analysis discloses a play upon such profound themes. This is why an understanding of rules of purity is a sound entry to comparative religion. The Pauline antithesis of blood and water, nature and grace, freedom and necessity, or the Old Testament idea of Godhead can be illuminated by Polynesian or Central African treatment of closely related themes.

I

Ritual Uncleanness

OUR IDEA OF DIRT is compounded of two things, care for hygiene and respect for conventions. The rules of hygiene change, of course, with changes in our state of knowledge. As for the conventional side of dirt-avoidance, these rules can be set aside for the sake of friendship. Hardy's farm labourers commended the shepherd who refused a clean mug for his cider as a 'nice unparticular man':

' "A clane cup for the shepherd," said the maltster commandingly.

' "No—not at all," said Gabriel, in a reproving tone of considerateness. "I never fuss about dirt in its pure state and when I know what sort it is. . . . I wouldn't think of giving such trouble to neighbours in washing up when there's so much work to be done in the world already." '

In a more exalted spirit, St. Catherine of Sienna, when she felt revulsion from the wounds she was tending, is said to have bitterly reproached herself. Sound hygiene was incompatible with charity, so she deliberately drank off a bowl of pus.

Whether they are rigorously observed or violated, there is nothing in our rules of cleanness to suggest any connection between dirt and sacredness. Therefore it is only mystifying to learn that primitives make little difference between sacredness and uncleanness.

For us sacred things and places are to be protected from defilement. Holiness and impurity are at opposite poles. We would as soon confound hunger with fullness or sleeping with waking. Yet it is supposed to be a mark of primitive religion to

7

make no clear distinction between sanctity and uncleanness. If this is true it reveals a great gulf between ourselves and our forefathers, between us and contemporary primitives. Certainly it has been very widely held and is still taught in one cryptic form or another to this day. Take the following remark of Eliade:

> 'The ambivalence of the sacred is not only in the psychological order (in that it attracts or repels), but also in the order of values; the sacred is at once "sacred" and "defiled".' (1958, p. 14-15)

The statement can be made to sound less paradoxical. It could mean that our idea of sanctity has become very specialised, and that in some primitive cultures the sacred is a very general idea meaning little more than prohibition. In that sense the universe is divided between things and actions which are subject to restriction and others which are not; among the restrictions some are intended to protect divinity from profanation, and others to protect the profane from the dangerous intrusion of divinity. Sacred rules are thus merely rules hedging divinity off, and uncleanness is the two-way danger of contact with divinity. The problem then resolves into a linguistic one, and the paradox is reduced by changing the vocabulary. This may be true of certain cultures. (See Steiner p. 33.)

For instance, the Latin word *sacer* itself has this meaning of restriction through pertaining to the gods. And in some cases it may apply to desecration as well as to consecration. Similarly the Hebrew root of k-d-sh, which is usually translated as Holy, is based on the idea of separation. Aware of the difficulty translating k-d-sh straight into Holy, Ronald Knox's version of the Old Testament uses 'set apart'. Thus the grand old lines 'Be ye Holy, Because I am Holy' are rather thinly rendered:

> 'I am the Lord your God, who rescued you from the land of Egypt; I am set apart and you must be set apart like me.'
> (Levit. 11.46)

If only re-translation could put the whole matter right, how simple it would be. But there are many more intractable cases. In Hinduism, for example, the idea that the unclean and the holy could both belong in a single broader linguistic category is ludicrous. But the Hindu ideas of pollution suggest another approach to the question. Holiness and unholiness after all need not always be absolute opposites. They can be relative categories. What is clean in relation to one thing may be unclean in relation to another, and vice versa. The idiom of pollution lends itself to a complex algebra which takes into account the variables in each context. For example, Professor Harper describes how respect can be expressed on these lines among the Havik peoples of the Malnad part of Mysore state:

> 'Behaviour that usually results in pollution is sometimes intentional in order to show deference and respect; by doing that which under other circumstances would be defiling, an individual expresses his inferior position. For example, the theme of the wife's subordination towards the husband finds ritual expression in her eating from his leaf after he has finished . . .'

In an even clearer case a holy woman, *sadhu*, when she visited the village, was required to be treated with immense respect. To show this the liquid in which her feet had been bathed:

> 'was passed round to those present in a special silver vessel used only for worshipping, and poured into the right hand to be drunk as *tirtha* (sacred liquid), indicating that she was being accorded the status of a god rather than a mortal. . . . The most striking and frequently encountered expression of respect-pollution is in the use of cow-dung as a cleansing agent. A cow is worshipped daily by Havik women and on certain ceremonial occasions by Havik men. . . . Cows are sometimes said to be gods; alternatively to have more than a thousand gods residing in them. Simple types of pollution are removed by water, greater degrees of pollution are removed by cow-dung and water. . . . Cow-dung, like the dung of any other animal, is intrinsically impure and can cause defilement—in fact it will defile a god; but it is pure relative to a mortal . . . the cow's most impure part is sufficiently pure relative even to a Brahmin priest to remove the latter's impurities.' (Harper, pp. 181-3)

It is obvious that we are here dealing with symbolic language capable of very fine degrees of differentiation. This use of the relation of purity and impurity is not incompatible with our own language and raises no specially puzzling paradoxes. So far from there being confusion between the idea of holiness and uncleanness, here there is nothing but distinction of the most hair-splitting finesse.

Eliade's statements about the confusion between sacred

contagion and uncleanness in primitive religion were evidently not intended to apply to refined Brahminical concepts. To what were they intended to apply? Apart from the anthropologists, are there any people who really confúse the sacred and the unclean? Where did this notion spring from?

Frazer seems to have thought that confusion between uncleanness and holiness is the distinctive mark of primitive thinking. In a long passage in which he considers the Syrian attitude to pigs, he concludes:

> 'Some said it was because pigs were unclean; others said it was because pigs were sacred. This . . . points to a hazy state of religious thought in which the idea of sanctity and uncleanness are not yet sharply distinguished, both being blent in a sort of vaporous solution to which we give the name taboo.'
>
> (Spirits of the Corn & Wild, II. p. 23)

Again he makes the same point in giving the meaning of taboo:

> 'Taboos of holiness agree with taboos of pollution because the savage does not distinguish between holiness and pollution.'
>
> (Taboo & the Perils of the Soul, p. 224)

Frazer had many good qualities, but originality was never one of them. These quotations directly echo Robertson Smith to whom he dedicated The Spirits of the Corn and Wild. Over twenty years earlier Robertson Smith had used the word taboo for restrictions on 'man's arbitrary use of natural things, enforced by dread of supernatural penalties' (1889, p. 142). These taboos, inspired by fear, precautions against malignant spirits, were common to all primitive peoples and often took the form of rules of uncleanness.

> 'The person under taboo is not regarded as holy, for he is separated from approach to the sanctuary, as well as from contact with men, but his act or condition is somehow associated with supernatural dangers, arising, according to the common savage explanation, from the presence of formidable spirits which are shunned like an infectious disease. In most savage societies no line seems to be drawn between the two kinds of taboo.'

According to this view the main difference between primitive taboo and primitive rules of holiness is the difference between friendly and unfriendly deities. The separation of sanctuary and

consecrated things and persons from profane ones, which is a
normal part of religious cults, is basically the same as the separations which are inspired by fear of malevolent spirits. Separation
is the essential idea in both contexts, only the motive is different
—and not so very different either, since friendly gods are also
to be feared on occasion. When Robertson Smith added that:—
'to distinguish between the holy and the unclean marks a real
advance above savagery', to his readers he was saying nothing
challenging or provocative. It was certain that his readers made
a big distinction between unclean and sacred, and that they
were living at the right end of the evolutionary movement. But
he was saying more than this. Primitive rules of uncleanness
pay attention to the material circumstances of an act and judge
it good or bad accordingly. Thus contact with corpses, blood or
spittle may be held to transmit danger. Christian rules of holiness, by contrast, disregard the material circumstances and judge
according to the motives and disposition of the agent.

> '. . . the irrationality of laws of uncleanness from the standpoint
> of spiritual religion or even of the higher heathenism, is so
> manifest that they must necessarily be looked upon as having
> survived from an earlier form of faith and of society.'
>
> (Note C. p. 430)

In this way a criterion was produced for classing religions as
advanced or as primitive. If primitive, then rules of holiness and
rules of uncleanness were indistinguishable; if advanced then
rules of uncleanness disappeared from religion. They were rele-
gated to the kitchen and bathroom and to municipal sanitation,
nothing to do with religion. The less uncleanness was concerned
with physical conditions and the more it signified a spiritual
state of unworthiness, so much more decisively could the religion
in question be recognised as advanced.

Robertson Smith was first and foremost a theologian and Old
Testament scholar. Since theology is concerned with the relations between man and God, it must always be making assertions
about the nature of man. At the time of Robertson Smith
anthropology was very much to the fore in theological discussion. Most thinking men in the second part of the nineteenth
century were perforce amateur anthropologists. This comes out
very clearly in Margaret Hodgen's *The Doctrine of Survivals*,
a necessary guide to the confused nineteenth century dialogue

between anthropology and theology. In that formative period anthropology still had its roots in the pulpit and parish hall, and bishops used its findings for fulminating texts.

Parish ethnologists took sides as pessimists or optimists on the prospects of human progress. Were the savages capable of advancement or not? John Wesley, teaching that mankind in its natural state was fundamentally bad, drew lively pictures of savage customs to illustrate the degeneracy of those who were not saved:

> 'The natural religion of the Creeks, Cherokees, Chickasaws and all other Indians, is to torture all their prisoners from morning to night, till at length they roast them to death. . . .
> Yea, it is a common thing among them for the son, if he thinks his father lives too long, to knock out his brains.'
> (Works, vol. 5, p. 402)

I need not here outline the long argument between the progressionists and degenerationists. For several decades the discussion dragged on inconclusively, until Archbishop Whately, in an extreme and popular form, took up the argument for degeneracy to refute the optimism of economists following Adam Smith.

> 'Could this abandoned creature,' he asked, 'entertain any of the elements of nobility? Could the lowest savages and the most highly civilised specimens of the European races be regarded as members of the same species? Was it conceivable as the great economist had asserted, that by the division of labour these shameless people could "advance step by step in all the arts of civilised life"?' (1855, pp. 26-7)

The reaction to his pamphlet, as Hodgen describes it, was intense and immediate:

> 'Other degenerationists, such as W. Cooke Taylor, composed volumes to support his position, assembling masses of evidence where the Archbishop had remained content with one illustration. . . . Defenders of the eighteenth century optimism appeared from all points of the compass. Books were reviewed in terms of Whately's contention. And social reformers everywhere, those good souls whose newly acquired compassion for the economically downtrodden had found a comfortable solvent in the notion of inevitable social improvement, viewed with alarm the practical outcome of the opposite view. . . . Even more disconcerted were those scholarly students of man's mind

and culture whose personal and professional interests were vested
in a methodology based upon the idea of progress.' (pp. 30-1)

One man finally came forward and settled the controversy for
the rest of the century by bringing science to the aid of the
progressionists. This was Henry Burnett Tylor (1832-1917). He
developed a theory and systematically amassed evidence to prove
that civilisation is the result of gradual progress from an original state similar to that of contemporary savagery.

> 'Among the evidence aiding us to trace the course which the
> civilisation of the world has actually followed is the great class
> of facts to denote which I have found it convenient to introduce
> the term "survivals". These are processes, customs, opinions and
> so forth, which have been carried by force of habit into the
> new society . . . and . . . thus remain as proofs and examples
> of an older condition of culture out of which a newer has
> evolved. (p. 16)
>   The serious business of ancient society may be seen to sink
> into the spirit of later generations and its serious belief to linger
> on in nursery folk-lore.' (p. 71)
>
> *(Primitive Culture I* 7th Edn.)

Robertson Smith used the idea of survivals to account for the
persistence of irrational rules of uncleanness. Tylor published in
1873, after the publication of the Origin of the Species, and there
is some parallel between his treatment of cultures and Darwin's
treatment of organic species. Darwin was interested in the conditions under which a new organism can appear. He was interested in the survival of the fittest and also in rudimentary
organs whose persistence gave him the clues for reconstruct-
ing the evolutionary scheme. But Tylor was uniquely interested
in the lingering survival of the unfit, in almost vanished cultural
relics. He was not concerned to catalogue distinct cultural species
or to show their adaptation through history. He only sought to
show the general continuity of human culture.

Robertson Smith, coming later, inherited the idea that modern
civilised man represents a long process of evolution. He accepted
that something of what we still do and believe is fossil; meaningless, petrified appendage to the daily business of living. But
Robertson Smith was not interested in dead survivals. Customs
which have not fed into the growing points of human history
he dubbed irrational and primitive and implied that they were
of little interest. For him the important task was to scrape away

the clinging rubble and dust of contemporary savage cultures and to reveal the life-bearing channels which prove their evolutionary status by their live functions in modern society. This is precisely what *The Religion of the Semites* attempts to do. Savage superstition is there separated from the beginnings of true religion, and discarded with very little consideration. What Robertson Smith says about superstition and magic is only incidental to his main theme and a by-product of his main work. Thus he reversed the emphasis of Tylor. Whereas Tylor was interested in what quaint relics can tell us of the past, Robertson Smith was interested in the common elements in modern and primitive experience. Tylor founded folk-lore: Robertson Smith founded social anthropology.

Another great stream of ideas impinged even more closely on Robertson Smith's professional interests. This was the crisis of faith which assailed those thinkers who could not reconcile the development of science with traditional Christian revelation. Faith and reason seemed hopelessly at odds unless some new formula for religion could be found. A group of philosophers who could no longer accept revealed religion, and who could not either accept to live without some guiding transcendental beliefs, set about providing that formula. Hence began that still-continuing process of whittling away the revealed elements of Christian doctrine, and the elevating in its place of ethical principles as the central core of true religion. In what follows I am quoting Richter's description of how the movement had its home in Oxford. At Balliol T. H. Green tried to naturalise Hegelian idealist philosophy as the solution to current problems of faith, morals and politics. Jowett had written to Florence Nightingale:

'Something needs to be done for the educated similar to what J. Wesley did for the poor.'

This is precisely what T. H. Green set out to achieve: to revive religion in the educated, make it intellectually respectable, create a new moral fervour and so produce a reformed society. His teaching had an enthusiastic reception. Complicated though his philosophic ideas were and tortuous their metaphysical basis, his principles were simple in themselves. They were even expressed in Mrs. Humphrey Ward's best-selling novel, *Robert Elsmere* (1888).

Green's philosophy of history was a theory of moral progress: God is made incarnate from age to age in social life of ever greater ethical perfection. To quote from his lay sermon—man's consciousness of God

'has in manifold forms been the moralising agent in human society, nay the formative principle of that society itself. The existence of specific duties and the recognition of them, the spirit of self-sacrifice, the moral law and reverence for it in its most abstract and absolute form, all no doubt presuppose society, but society of a kind to render them possible is not the creature of appetite and fear. . . . Under this influence wants and desires that have their root in the animal nature become an impulse of improvement which forms, enlarges and recasts societies, always keeping before man in various guises according to the degree of his development an unrealised ideal of a best which is his God, and giving divine authority to the customs or laws by which some likeness of this ideal is wrought into the actuality of life.' (Richter, p. 105)

The final trend of Green's philosophy was thus to turn away from revelation and to enshrine morality as the essence of religion. Robertson Smith never turned away from Revelation. To the end of his life he believed in the divine inspiration of the Old Testament. His biography by Black and Chrystal suggests that in spite of this belief he came strangely close to the Oxford Idealists' notion of religion.

Robertson Smith held the Free Church Chair of Hebrew in Aberdeen in 1870. He was in the vanguard of the movement of historical criticism which for some time earlier had been making upheavals in the conscience of Biblical scholars. In 1860 Jowett himself at Balliol had been censured for publishing an article 'On the Interpretation of the Bible', in which he argued that the Old Testament must be interpreted like any other book. Proceedings against Jowett collapsed and he was allowed to remain Regius Professor. But when Robertson Smith wrote the article, Bible, in 1875, for the Encyclopaedia Britannica, the outcry in the Free Church against his heresy led to his suspension and dismissal. Robertson Smith, like Green, was in close touch with German thought, but whereas Green was not committed to Christian revelation, Robertson Smith never wavered in his faith in the Bible as the record of a specific, supernatural Revelation. Not only was he prepared to treat its books to the same kind of criticism as other books, but after he was dismissed from Aberdeen he travelled in Syria and brought informed field-work to its interpretation. On the basis of this first-hand study of Semitic life and documents he delivered the Burnett lectures. The first series of these was published as *The Religion of the Semites*.

From the way he wrote it is clear that this study was no ivory-tower escape from the real problems of humanity of his day. It was important to understand the religious beliefs of obscure Arab tribes because these shed light on the nature of man and on the nature of religious experience. Two important themes emerged from his lectures. One is that exotic mythological happenings and cosmological theories had little to do with religion. Here he is implicitly contradicting Tylor's theory that primitive religion arose from speculative thought. Robertson Smith suggested that those who were lying awake at night trying to reconcile the details of the Creation in Genesis with the Darwinian theory of evolution could relax. Mythology is so much extra embroidery on more solid beliefs. True religion, even from the earliest times, is firmly rooted in the ethical values of community life. Even the most misguided primitive neighbours of Israel, bedevilled by demons and myths, still showed some signs of true religion.

The second theme was that Israel's religious life was funda-mentally more ethical than that of any of the surrounding peoples. Let us take this second point quickly first. The last three Burnett lectures, given in Aberdeen in 1891, were not published and little now survives of them. The lectures dealt with apparent Semitic parallels with the cosmogony of Genesis. The alleged parallel with the Chaldean cosmogony was held by Robert-son Smith to have been much exaggerated, and the Baby-lonian myths were classed by him as more like the myths of savage nations than those of Israel. The Phoenician legend, again, superficially resembles the Genesis story, but the similarities serve to bring out the deep differences of spirit and meaning:

'Phoenician legends . . . were bound up with a thoroughly heathen view of God, man and the world. Destitute as these legends were of ethical motives, no believer in them could rise to any spiritual conception of Deity nor any lofty conception of man's chief end . . . The burden of explaining this contrast (with Hebrew ideas of deity) does not lie with me. It falls on those who are compelled by a false philosophy of Revelation to see in the Old Testament nothing more than the highest point of the general tendencies of Semitic religions. This is not the view that study commends to me. It is a view that is not commended, but condemned by the many parallelisms in detail between Hebrew and heathen story and ritual, for all these material points of resemblance only make the contrast in spirit the more remarkable. . . .' [Black & Chrystal, p. 536]

So much for the overwhelming inferiority of the religion of Israel's neighbours, and heathen Semites. As for the basis of heathen Semitic religions, it has two characteristics: an abounding demonology, rousing fear in men's hearts, and a comforting, stable relation with the community god. The demons are the primitive element rejected by Israel; the stable, moral relation with God is true religion.

'However true it is that savage man feels himself to be environed by innumerable dangers which he does not understand and so personifies as invisible or mysterious enemies of more than human power, it is not true that the attempt to appease these powers is the foundation of religion. From the earliest times religion, as distinct from magic and sorcery, addressed itself to kindred and friends who may indeed be angry with their people for a time, but are always placable except to the enemies of their people or to renegade members of the community. . . . It is only in times of social dissolution . . . that magical superstition based on mere terror or rites designed to placate alien gods invade the sphere of tribal or national religion. In better times the religion of the tribe or state has nothing in common with the private and foreign superstitions or magical rites that savage terror may dictate to the individual. Religion is not an arbitrary relation of the individual man to a supernatural power; it is a relation of all the members of a community to the power that has the good of the community at heart.' (Religion of the Semites p. 55)

It is clear that in the 1890's this authoritative pronouncement on the relation of morals to primitive religion would be warmly welcomed. It would bring together in a happy combination the new ethical idealism of Oxford and ancient revelation. It is plain that Robertson Smith himself had fallen entirely for the ethical view of religion. The compatibility of his views with those advanced in Oxford is nicely confirmed in the fact that when he was first dismissed from the Chair of Hebrew in Aberdeen, Balliol offered him a post.

He was confident that the pre-eminence of the Old Testament would stand above the challenge however close the scientific scrutiny. For he could show with unrivalled erudition that all primitive religions express social forms and values. And since the moral loftiness of Israel's religious concepts was above dispute, and since these had given way in the course of history to the ideals of Christianity and these in turn had moved from Catholic to Protestant forms, the evolutionary movement was clear. Science was thus not opposed but deftly harnessed to the Christian's task.

From this point onwards the anthropologists have been saddled with an intractable problem. For magic is defined for them in residual, evolutionary terms. In the first place it is ritual which is not part of the cult of the community's god. In the second place it is ritual which is expected to have automatic effect. In a sense magic was to the Hebrews what Catholicism was to the Protestants, mumbo-jumbo, meaningless ritual, irrationally held to be sufficient in itself to produce results without an interior experience of God.

Robertson Smith in his inaugural lecture contrasts the intelligent, Calvinist approach with the magical treatment of the Scriptures practised by the Roman Catholics who loaded the Book with superstitious accretions. In the same inaugural lecture he drives home the point. 'The Catholic Church', he argued:

'had almost from the first deserted the Apostolic tradition and set up a conception of Christianity as a mere series of formulae containing abstract and immutable principles, intellectual assent to which was sufficient to mould the lives of men who had no experience of a personal relation with Christ. . . .

Holy Scripture is not, as the Catholics tend to claim, "a divine phenomenon magically endowed in every letter with saving treasures of faith and knowledge".' (Black & Chrystal, pp. 126-7)

His biographers suggest that the association of magic with Catholicism was a canny move to shame his die-hard Protestant opponents into more courageous intellectual dealings with the Bible. Whatever the Scot's motives the historical fact remains that comparative religion has inherited an ancient sectarian quarrel about the value of formal ritual. And now the time has come to show how an emotional and prejudiced approach to ritual has led anthropology down one of its barrenest perspectives—a narrow preoccupation with belief in the efficacy of rites. This I shall develop in Chapter 4. While Robertson Smith was perfectly right to recognise in the history of Christianity an ever-present tendency to slip into purely formal and instrumental use of ritual, his evolutionary assumptions misled him twice. Magical practice, in this sense of automatically effective ritual, is not a sign of primitiveness, as the contrast he himself drew between the religion of the Apostles and that of late Catholicism should have suggested. Nor is a high ethical content the prerogative of evolved religions, as I hope to show in later chapters.

The influence which Robertson Smith exerted divides into two streams according to the uses to which Durkheim and Frazer put his work. Durkheim took up his central thesis and set comparative religion on fruitful lines. Frazer took up his incidental minor thesis, and sent comparative religion into a blind alley.

Durkheim's debt to Robertson-Smith is acknowledged in the *Elementary Forms of the Religious Life* (p. 61). His whole book develops the germinal idea that primitive gods are part and parcel of the community, their form expressing accurately the details of its structure, their powers punishing and rewarding on its behalf. In primitive life:

'Religion was made up of a series of acts and observances, the correct performance of which was necessary or desirable to secure the favour of the gods or to avert their anger, and in their observances every member of society had a share marked out for him either in virtue of being born within the family and community or in virtue of the station within the family and community that he had come to hold . . . Religion did not exist for the saving of souls but for the preservation and welfare of society . . . A man was born into a fixed relation with certain gods as surely as he was born into a relation with his fellow men; and his religion, that is the part of conduct which was determined by his relation to the gods, was simply one side of the general scheme of conduct prescribed for him by his position as a member of society . . . Ancient religion is but part of the general social order which embraces gods and men alike.'

Thus wrote Robertson Smith (pp. 29-33). But for differences of style and the use of the past tense it could have been written by Durkheim.

I find it very helpful to understand Durkheim as engaged initially in an argument with the English, as Talcott Parsons has suggested (1960). He was concerned with a particular problem about social integration posed for him by the shortcomings of English political philosophy, particularly represented by Herbert Spencer. He could not subscribe to the utilitarian theory that the psychology of the individual would account for the development of society. Durkheim wanted to show that something else was necessary, a common commitment to a common set of values, a collective conscience, if the nature of society was to be correctly understood. At the same time another Frenchman, Gustav le Bon (1841-1931) was engaged also in the same task of correcting the prevailing Benthamite tradition. He proceeded by developing a theory of crowd psychology which Durkheim also seems to have freely drawn upon. Compare Durkheim's account of the emotional force of totemic ceremonies (p. 241) with le Bon's account of the suggestible, emotionally savage or heroic 'crowd mind'. But a better instrument to Durkheim's purpose of convicting the English of error, was the work of another Englishman.

Durkheim adopted in its entirety Robertson Smith's definition of primitive religion as the established church which expresses community values. He also followed Robertson Smith unquestioningly in his attitude to rites which were not part of the cult of the community gods. He followed him in dubbing these 'magic' and defined magic and magicians as beliefs, practices and persons not operating within the communion of the church and often hostile to it. Following Robertson Smith and perhaps following Frazer, the early volumes of whose *Golden Bough* were already published when the *Elementary Forms of Religious Life* came out in 1912, he allowed that magic rites were a form of primitive hygiene:

> 'The things which the magician recommends to be kept separate are those which, by reason of their characteristic properties, cannot be brought together and confused without danger . . . useful maxims, the first forms of hygienic and medical interdictions.' (p. 338)

Thus the distinction between contagion and true religion was confirmed. Rules of uncleanness lay outside the main stream of his interests. He paid them no more attention than did Robertson Smith.

But any arbitrary limitation of his subject draws a scholar into difficulty. When Durkheim set aside one class of separations as primitive hygiene and another class as primitive religion he undermined his own definition of religion. His opening chapters summarise and reject unsatisfactory definitions of religion. Attempts to define religion by notions of mystery and awe he dismisses, and likewise Tylor's definition of religion as belief in spiritual beings. He proceeds to adopt two criteria which he assumes will be found to coincide; the first, we have seen, is the communal organisation of men for the community cult, and the second is the separation of sacred from profane. The sacred is the object of community worship. It can be recognised by rules expressing its essentially contagious character.

In insisting on a complete break between the sphere of the sacred and the sphere of the profane, between secular and religious behaviour, Durkheim is not following in Robertson Smith's footsteps. For the latter took the opposite view and insisted (p. 29 seq.) that there is 'no separation between the spheres of religion and ordinary life'. A total opposition between sacred and profane seems to have been a necessary step in Durkheim's theory of social integration. It expressed the opposition between the individual and society. The social conscience was projected beyond and above the individual member of society onto something quite other, external and compellingly powerful. So we find Durkheim insisting that rules of separation are the distinguishing marks of the sacred, the polar opposite of the profane. He then is led by his argument into asking why the sacred should be contagious. This he answers by reference to the fictive, abstract nature of religious entities. They are merely ideas awakened by the experience of society, merely collective ideas projected outwards, mere expressions of morality. So they have no fixed material point of reference. Even the graven images of gods are only material emblems of immaterial forces generated by the social process. Therefore they are ultimately rootless, fluid, liable to become unfocussed and to flow into other experiences. It is their nature always to be in danger of losing their distinctive and necessary character. The sacred needs to be continually hedged in with prohibitions. The sacred must always be treated as contagious because relations with it are bound to be expressed by rituals of separation and demarcation and by beliefs in the danger of crossing forbidden boundaries.

There is one little difficulty about this approach. If the sacred is characterised by its contagiousness, how does it differ from non-sacred magic, also characterised by contagiousness? What is the status of the other kind of contagiousness which is not generated from the social process? Why are magical beliefs called primitive hygiene and not primitive religion? These problems did not interest Durkheim. He followed Robertson Smith in cutting off magic from morals and religion and so helped to bequeath to us a tangle of ideas about magic. Ever since scholars have scratched their heads for a satisfactory definition of magic beliefs and then puzzled over the mentality of people who can subscribe to them.

It is easy now to see that Durkheim advocated an altogether too unitary view of the social community. We should start by recognising communal life for a much more complex experience than he allowed. Then we find that Durkheim's idea of ritual as symbolic of social processes, can be extended to include both types of belief in contagion, religious and magical. If he could have foreseen an analysis of ritual in which none of the rules which he called hygienic are without their load of social symbolism, he would presumably have been happy to discard the category of magic. To this theme I shall return. But we cannot develop it without first rubbing the slate clean of another set of preconceptions which derive also from Robertson Smith.

Frazer was not interested in the sociological implications of Robertson Smith's work. He seems indeed not to have been very interested in its main theme at all. Instead he fastened on the magical residue which was thrown off incidentally, as it were, from the definition of true religion. He showed that there were certain regularities to be found in magical beliefs and that these could be classified. On inspection magic turned out to be much more than mere rules of avoiding obscure infection. Some magic acts were intended to procure benefits and others to avert harm. So the field of behaviour which Robertson Smith labelled superstition held more than rules of uncleanness. But contagion seemed to be one of its governing principles. The other principle was belief in the transfer of properties by sympathy or likeness.

According to the so-called laws of magic the magician can change events either by mimetic action or by allowing contagious forces to work. When he had finished investigating magic Frazer had done no more than to name the conditions under which one thing may symbolise another. If he had not been convinced that savages think on entirely different lines from ourselves, he might have been content to treat magic as symbolic action, neither more nor less. He might then have joined hands with Durkheim and the French school of sociology and the dialogue across the channel would have been more fruitful for English nineteenth century thought. Instead he crudely tidied up the evolutionary assumptions implicit in Robertson Smith and assigned to human culture three stages of development.

Magic was the first stage, religion the second, science the third. His argument proceeds by a kind of Hegelian dialectic since magic, classed as primitive science, was defeated by its own inadequacy and supplemented by religion in the form of a priestly and political fraud. From the thesis of magic emerged the antithesis, religion, and the synthesis, modern effective science, replaced both magic and religion. This fashionable presentation was supported by no evidence whatever. Frazer's evolutionary scheme was only based on some unquestioning assumptions taken over from the common talk of his day. One was the assumption that ethical refinement is a mark of advanced civilisation. Another the assumption that magic has nothing to do with morals or religion. On this basis he constructed the image of our early ancestors, their thinking dominated by magic. For them the universe was moved by impersonal, mechanistic principles. Fumbling for the right formula for controlling it, they stumbled on some sound principles, but just as often their confused state of mind led them to think that words and signs could be used as instruments. Magic resulted from early man's inability to distinguish between his own subjective associations and external objective reality. Its origin was based on a mistake. No doubt about it, the savage was a credulous fool.

'Thus the ceremonies which in many lands have been performed to hasten the departure of winter or stay the flight of summer are in a sense an attempt to create the world afresh, to "remould it nearer to the heart's desire". But if we would set ourselves at the point of view of the old sages who devised means so feeble

23

to accomplish a purpose so immeasurably vast, we must divest ourselves of our modern conceptions of the immensity of the Universe and of the pettiness and insignificance of man's place in it . . . To the savage the mountains that bound the visible horizon, or the sea that stretches away to meet it, is the world's end. Beyond these narrow limits his feet have never strayed . . . of the future he hardly thinks, and of the past he only knows what has been handed down by word of mouth from his savage forefathers. To suppose that a world thus circumscribed in space and time was created by the efforts or the fiat of a being like himself imposes no great strain on his credulity; and he may without much difficulty imagine that he himself can annually repeat the work of creation by his charms and incantations.' (Spirits of the Corn and Wild, II, p. 109).

It is hard to forgive Frazer for his complacency and undisguised contempt of primitive society. The last chapter of Taboo and the Perils of the Soul is entitled, 'Our Debt to the Savage'. Possibly it was inserted in response to correspondents who pressed him to recognise the wisdom and philosophic depth of primitive cultures they knew. Frazer gives interesting extracts from these letters in footnotes, but he cannot adjust his own preconceived judgments to take them into account. The chapter purports to contain a tribute to savage philosophy, but since Frazer could offer no reason for respecting ideas which he had massively demonstrated to be childish, irrational and superstitious, the tribute is mere lip service. For pompous patronage this is hard to beat:

'When all is said and done, our resemblances to the savage are still far more numerous than our differences . . . after all, what we call truth is only the hypothesis which is found to work best. Therefore in reviewing the opinions and practices of ruder ages and races we shall do well to look with leniency upon their errors as inevitable slips made in the search for truth. . . .'

Frazer had his critics and they gained some attention at the time. But in England Frazer undoubtedly triumphed. For is not the *Golden Bough* abridged edition still bringing in an income? Is not a Frazer Memorial Lecture regularly delivered? Partly it was the very simplicity of his views, partly the tireless energy which brought out volume after volume, but above all the gold and purple style of writing, which gave such wide circulation to his work. In almost any study of ancient civilisations you will be sure to find continual references to primitiveness and to its criterion, magical non-ethical superstition.

Take Cassirer, for example, writing about Zoroastrianism, and recognise these themes from the *Golden Bough*:

> 'Even nature assumes a new shape, for it is seen exclusively in the mirror of ethical life. Nature . . . is conceived as the sphere of law and lawfulness. In Zoroastrian religion nature is described by the concept of *Asha*. *Asha* is the wisdom of nature that reflects the wisdom of its creator, of Ahura Mazda, the "wise Lord". This universal, eternal, inviolable order governs the world and determines all single events: the path of the sun, the moon, the stars, the growth of plants and animals, the way of winds and clouds. All this is maintained and preserved, not by mere physical forces but by the force of the Good . . . the ethical meaning has replaced and superseded the magical meaning.'
> (1944, p. 100)

Or to take a more recent source on the same subject, we find Professor Zaehner noting sadly that the least defective Zoroastrian texts are only concerned with rules of purity and therefore of no interest:

> '. . . only in the Vidēvdāt with its dreary prescriptions concerning ritual purity and its listing of impossible punishments for ludicrous crimes do the translators show a tolerable grasp of the text.'
> (pp. 25-6)

This is certainly how Robertson Smith would consider such rules, but 70 years later can we be sure that this is all there is to be said about them?

In Old Testament studies the assumption is rife that primitive peoples use rituals magically, that is in a mechanical, instrumental way. 'In early Israel the distinction between what we call intentional and unintentional sin, as far as God is concerned, scarcely exists' (Osterley & Box). 'For the Hebrews of the fifth century B.C.', writes Professor James, 'expiation was merely a mechanical process consisting of wiping away material uncleanness' (1938). The history of the Israelites is sometimes presented as a struggle between the prophets who demanded interior union with God and the people, continually liable to slide back into primitive magicality, to which they are particularly prone when in contact with other more primitive cultures. The paradox is that magicality seems finally to prevail with the compilation of the Priestly Code. If belief in the sufficient efficacy of the rite is to be called magic in its late as well as in its earliest manifestations, the usefulness of magic as a measure of primitiveness would be lost. One would expect the very word to be expunged from Old Testament studies. But it lingers on, with *Tabu* and *mana*, to emphasise the distinctiveness of the Israelite religious experience by contrast with Semitic heathenism. Eichrodt is particularly free with these terms (pp. 438, 453):

> 'Mention has already been made of the magical effect ascribed to Babylonian rites and formulas of expiation, and this becomes especially clear when it is remembered that the confession of sin actually forms part of the ritual of exorcism and has *ex opere operato* efficiency.'
>
> (p. 166)

He goes on to cite Psalms 40, 7, and 69, 31, as 'opposing the tendency of the sacrificial system to make forgiveness of sins a mechanical process'. Again on p. 119 he assumes that primitive religious concepts are 'materialistic'. Much of this otherwise impressive book rests on the assumption that ritual which works *ex opere operato* is primitive, prior in time compared with ritual which is symbolic of internal states of mind. But occasionally the unattested *a priori* nature of this assumption seems to make the author uneasy:

> 'The commonest of all expressions for making atonement, *kipper*, also points in this direction if the original meaning of the term may be defined as "to wipe away" on the basis of the Babylonian and Assyrian parallels. Here the fundamental concept of sin is of a material impurity, and the blood, as a holy substance endowed with miraculous power, is expected to remove the stain of sin quite automatically.'

Then comes an illumination which would cause much rewriting if taken seriously:

> 'Since, however, the derivation based on the Arabic, giving the meaning "to cover" seems equally possible, it may well be that the idea is that of covering up one's guilt from the eyes of the offended party by means of reparation, which would by contrast emphasise the personal character of the act of atonement.'
>
> (p. 162)

So Eichrodt half relents towards the Babylonians—perhaps they too knew something of true interior religion; perhaps the

26

Israelite religious experience did not stick out in the surrounding pagan magic with such unique distinctiveness.

We find some of the same assumptions governing the interpretation of Greek literature. Professor Finley, in discussing the social life and beliefs of Homer's world, applies an ethical test for distinguishing earlier from later elements of belief (pp. 147, 151, 157).

Again, a learned French classicist, Moulinier, makes a comprehensive study of ideas of purity and impurity in Greek thought. Free of the bias of Robertson Smith, his approach seems excellently empirical by current anthropological standards. Greek thought seems to have been relatively free of ritual pollution in the period which Homer describes (if there was such a historical period), while clusters of pollution concepts emerge later and are expressed by the classical dramatists. The anthropologist, weak in classical scholarship, looks round for specialist guidance on how much reliance can be placed in this author, for his material is challenging and, to the layman, convincing. Alas —the book is roundly condemned in the Journal of Hellenic Studies by an English reviewer who finds it wanting in nineteenth-century anthropology:

'. . . the author needlessly handicapped himself. He appears to know nothing of the great mass of comparative material which is available to anyone studying purity, pollution and purification . . . a very modest amount of anthropological knowledge would tell him that so old a notion as that of pollution of shed blood belongs to a time when the community was the whole world . . . on p. 277 he uses the word "tabu" but only to show that he has no clear idea of what it means.' (Rose, 1954)

Whereas a reviewer unburdened by dubious anthropological knowledge recommends Moulinier's work without reserve (Whatmough).

These scattered quotations collected very much at random could easily be multiplied. They show how widespread Frazer's influence has been. Within anthropology too, his work has gone very deep. It seems that once Frazer had said that the interesting question in comparative religion hinged on false beliefs in magical efficacy, British anthropologists' heads remained dutifully bowed over this perplexing question, even though they had long rejected the evolutionary hypotheses which for Frazer made it interesting. So we read through virtuoso displays of learning on the relation between magic and science whose theoretical importance remains obscure.

All in all, Frazer's influence has been a baneful one. He took from Robertson Smith that scholar's most peripheral teaching, and perpetuated an ill-considered division between religion and magic. He disseminated a false assumption about the primitive view of the universe worked by mechanical symbols, and another false assumption that ethics are strange to primitive religion. Before we can approach the subject of ritual defilement these assumptions need to be corrected. The more intractable puzzles in comparative religion arise because human experience has been thus wrongly divided. In this book we try to re-unite some of the separated segments.

In the first place we shall not expect to understand religion if we confine ourselves to considering belief in spiritual beings, however the formula may be refined. There may be contexts of enquiry in which we should want to line up all extant beliefs in other beings, zombies, ancestors, demons, fairies—the lot. But following Robertson Smith we should not suppose that in cataloguing the full spiritual population of the universe we have necessarily caught the essentials of religion. Rather than stopping to chop definitions, we should try to compare peoples' views about man's destiny and place in the universe. In the second place we shall not expect to understand other people's ideas of contagion, sacred or secular, until we have confronted our own.

2

Secular Defilement

COMPARATIVE RELIGION has always been bedevilled by medical materialism. Some argue that even the most exotic of ancient rites have a sound hygienic basis. Others, though agreeing that primitive ritual has hygiene for its object, take the opposite view of its soundness. For them a great gulf divides our sound ideas of hygiene from the primitive's erroneous fancies. But both these medical approaches to ritual are fruitless because of a failure to confront our own ideas of hygiene and dirt.

On the first approach it is implied that if we only knew all the circumstances we would find the rational basis of primitive ritual amply justified. As an interpretation this line of thought is deliberately prosaic. The importance of incense is not that it symbolises the ascending smoke of sacrifice, but it is a means of making tolerable the smells of unwashed humanity. Jewish and Islamic avoidance of pork is explained as due to the dangers of eating pig in hot climates.

It is true that there can be a marvellous correspondence between the avoidance of contagious disease and ritual avoidance. The washings and separations which serve the one practical purpose may be apt to express religious themes at the same time. So it has been argued that their rule of washing before eating may have given the Jews immunity in plagues. But it is one thing to point out the side benefits of ritual actions, and another thing to be content with using the by-products as a sufficient explanation. Even if some of Moses's dietary rules were hygienically beneficial it is a pity to treat him as an enlightened public health administrator, rather than as a spiritual leader.

I quote from a commentary on Mosaic dietary rules, dated 1841:

'. . . It is probable that the chief principle determining the laws of this chapter will be found in the region of hygiene and sanitation. . . . The idea of parasitic and infectious maladies, which has conquered so great a position in modern pathology, appears to have greatly occupied the mind of Moses, and to have dominated all his hygienic rules. He excludes from the Hebrew dietary animals particularly liable to parasites; and as it is in the blood that the germ or spores of infectious diseases circulate, he orders that they must be drained of their blood before serving for food. . . .'

(Kellog)

He goes on to quote evidence that European Jews have a longer expectation of life and immunity in plagues, advantages which he attributes to their dietary restrictions. When he writes of parasites, it is unlikely that Kellog is thinking of the trichiniasis worm, since it was not observed until 1828 and was considered harmless to man until 1860 (Hegner, Root and Augustine, 1924, p. 439).

For a recent expression of the same kind of view read Dr. Ajose's account of the medical value of ancient Nigerian practices (1957). The Yoruba cult of a smallpox deity, for example, requires the patients to be isolated and treated only by a priest, himself recovered from the disease and therefore immune. Furthermore, the Yoruba use the left hand for handling anything dirty,

'because the right hand is used for eating, and people realise the risk of contamination of food that might result if this distinction were not observed.'

Father Lagrange also subscribed to the same idea:

'*Alors l'impurité, nous ne le nions pas, a un caractère religieux, ou du moins touche au surnaturel prétendu; mais, dans sa racine, est-ce autre chose qu'une mesure de préservation sanitaire? L'eau ne remplace-t-elle pas ici les antiseptiques? Et l'esprit redouté n'a-t-il pas fait des siennes en sa nature propre de microbe?*'

(p. 155)

It may well be that the ancient tradition of the Israelites included the knowledge that pigs are dangerous food for humans. Anything is possible. But note that this is not the reason given in Leviticus for the prohibition of pork and evidently the tradition, if it ever existed, was lost. For Maimonides himself, the great twelfth-century prototype of medical materialism, although he could find hygienic reasons for all the other dietary restrictions of Mosaic law, confessed himself baffled by the prohibition on pork, and was driven back to aesthetic explanations, based on the revolting diet of the domestic pig:

'I maintain that the food which is forbidden by the Law is unwholesome. There is nothing among the forbidden kinds of food whose injurious character is doubted, except pork, and fat. But also in these cases the doubt is not justified. For pork contains more moisture than necessary (for human food), and too much of superfluous matter. The principal reason why the Law forbids swine's flesh is to be found in the circumstance that its habits and its food are very dirty and loathsome. . . .'

(p. 370 seq.)

This at least shows that the original basis of the rule concerning pig flesh was not transmitted with the rest of the cultural heritage, even if it had once been recognised.

Pharmacologists are still hard at work on Leviticus XI. To give one example I cite a report by David I. Macht to which Miss Jocelyne Richard has referred me. Macht made muscle extract from swine, dog, hare, coney (equated with guinea-pigs for experimental purposes), and camel, and also from birds of prey and from fishes without fins and scales. He tested the extracts for toxic juices and found them to be toxic. He tested extracts from animals which counted as clean in Leviticus and found them less toxic, but still he reckoned his research proved nothing either way about the medical value of the Mosaic laws.

For another example of medical materialism read Professor Kramer, who lauds a Sumerian tablet from Nippur as the only medical text received from the 3rd millenium B.C.

'The text reveals, though indirectly, a broad acquaintance with quite a number of rather elaborate medical operations and procedures. For example, in several of the prescriptions the instructions were to "purify" the simples before pulverisation, a step which must have required several chemical operations.'

Quite convinced that purifying here does not mean sprinkling with holy water or reciting a spell, he goes on enthusiastically:

'The Sumerian physician who wrote our tablet did not resort to magic spells and incantations . . . the startling fact remains that our clay document, the oldest "page" of medical text as yet uncovered, is completely free from mystical and irrational elements.'
(1956, pp. 58-9)

So much for medical materialism, a term coined by William James for the tendency to account for religious experience in these terms: for instance, a vision or dream is explained as due to drugs or indigestion. There is no objection to this approach unless it excludes other interpretations. Most primitive peoples are medical materialists in an extended sense, in so far as they tend to justify their ritual actions in terms of aches and pains which would afflict them should the rites be neglected. I shall later show why ritual rules are so often supported with beliefs that specific dangers attend on their breach. By the time I have finished with ritual danger I think no one should be tempted to take such beliefs at face value.

As to the opposite view—that primitive ritual has nothing whatever in common with our ideas of cleanness—this I deplore as equally harmful to the understanding of ritual. On this view our washing, scrubbing, isolating and disinfecting has only a superficial resemblance with ritual purifications. Our practices are solidly based on hygiene; theirs are symbolic: we kill germs, they ward off spirits. This sounds straightforward enough as a contrast. Yet the resemblance between some of their symbolic rites and our hygiene is sometimes uncannily close. For example, Professor Harper summarises the frankly religious context of Havik Brahmin pollution rules. They recognise three degrees of religious purity. The highest is necessary for performing an act of worship; a middle degree is the expected normal condition, and finally there is a state of impurity. Contact with a person in the middle state will cause a person in the highest state to become impure, and contact with anyone in an impure state will make either higher categories impure. The highest state is only gained by a rite of bathing.

'A daily bath is absolutely essential to a Brahmin, for without it he cannot perform daily worship to his gods. Ideally, Haviks say, they should take three baths a day, one before each meal. But few do this. In practice all Haviks whom I have known rigidly observe the custom of a daily bath, which is taken before the main meal of the day and before the household gods are

worshipped. . . . Havik males, who belong to a relatively
wealthy caste and who have a fair amount of leisure time during
certain seasons, nevertheless do a great deal of the work required
to run their areca nut estates. Every attempt is made to finish
work that is considered dirty or ritually defiling—for example,
carrying manure to the garden or working with an untouchable
servant—before the daily bath that precedes the main meal.
If for any reason this work has to be done in the afternoon,
another bath should be taken when the man returns home. . . .'

(p. 153)

A distinction is made between cooked and uncooked food as
carriers of pollution. Cooked food is liable to pass on pollution,
while uncooked food is not. So uncooked foods may be received
from or handled by members of any caste—a necessary rule
from the practical point of view in a society where the division
of labour is correlated with degrees of inherited purity. (See
p. 127 in Chapter 7.) Fruit and nuts, as long as they are whole, are
not subject to ritual defilement, but once a coconut is broken or
a plantain cut, a Havik cannot accept it from a member of a
lower caste.

'The process of eating is potentially polluting, but the manner
determines the amount of pollution. Saliva—even one's own—is
extremely defiling. If a Brahmin inadvertently touches his
fingers to his lips, he should bathe or at least change his clothes.
Also, saliva pollution can be transmitted through some material
substances. These two beliefs have led to the practice of drinking water by pouring it into the mouth instead of putting the
lips on the edge of the cup, and of smoking cigarettes . . .
through the hand so that they never directly touch the lips.
(Hookas are virtually unknown in this part of India) . . . Eating
of any food—even drinking coffee—should be preceded by washing the hands and feet.'

(p. 156)

Food which can be tossed into the mouth is less liable to convey
saliva pollution to the eater than food which is bitten into. A
cook may not taste the food she is preparing, as by touching her
fingers to her lips she would lose the condition of purity required
for protecting food from pollution. While eating a person is in
the middle state of purity and if by accident he should touch
the server's hand or spoon, the server becomes impure and
should at least change clothes before serving more food. Since
pollution is transmitted by sitting in the same row at a meal,

33

when someone of another caste is entertained he is normally seated separately. A Havik in a condition of grave impurity should be fed outside the house, and he is expected himself to remove the leaf-plate he fed from. No one else can touch it without being defiled. The only person who is not defiled by touch and by eating from the leaf of another is the wife who thus, as we have said, expresses her personal relation to her husband. And so the rules multiply. They discriminate in ever finer and finer divisions, prescribing ritual behaviour concerning menstruation, childbirth and death. All bodily emissions, even blood or pus from a wound, are sources of impurity. Water, not paper must be used for washing after defaecating, and this is done only with the left hand, while food may be eaten only with the right hand. To step on animal faeces causes impurity. Contact with leather causes impurity. If leather sandals are worn they should not be touched with the hands, and should be removed and the feet be washed before a temple or house is entered.

Precise regulations give the kinds of indirect contact which may carry pollution. A Havik, working with his untouchable servant in his garden, may become severely defiled by touching a rope or bamboo at the same time as the servant. It is the simultaneous contact with the bamboo or rope which defiles. A Havik cannot receive fruit or money directly from an Untouchable. But some objects stay impure and can be conductors of impurity even after contact. Pollution lingers in cotton cloth, metal cooking vessels, cooked food. Luckily for collaboration between the castes, ground does not act as a conductor. But straw which covers it does.

> 'A Brahmin should not be in the same part of his cattle shed as his Untouchable servant, for fear that they may both step on places connected through overlapping straws on the floor. Even though a Havik and an Untouchable simultaneously bathe in the village pond, the Havik is able to attain a state of *Madi* (purity) because the water goes to the ground, and the ground does not transmit impurity.'
>
> (p. 173)

The more deeply we go into this and similar rules, the more obvious it becomes that we are studying symbolic systems. Is this then really the difference between ritual pollution and our ideas of dirt: Are our ideas hygienic where theirs are symbolic?

Not a bit of it: I am going to argue that our ideas of dirt also express symbolic systems and that the difference between pollution behaviour in one part of the world and another is only a matter of detail.

Before we start to think about ritual pollution we must go down in sack-cloth and ashes and scrupulously re-examine our own ideas of dirt. Dividing them into their parts, we should distinguish any elements which we know to be the result of our recent history.

There are two notable differences between our contemporary European ideas of defilement and those, say, of primitive cultures. One is that dirt avoidance for us is a matter of hygiene or aesthetics and is not related to our religion. I shall say more about the specialisation of ideas which separates our notions of dirt from religion in Chapter 5 (Primitive Worlds). The second difference is that our idea of dirt is dominated by the knowledge of pathogenic organisms. The bacterial transmission of disease was a great nineteenth-century discovery. It produced the most radical revolution in the history of medicine. So much has it transformed our lives that it is difficult to think of dirt except in the context of pathogenicity. Yet obviously our ideas of dirt are not so recent. We must be able to make the effort to think back beyond the last 100 years and to analyse the bases of dirt-avoidance, before it was transformed by bacteriology; for example, before spitting deftly into a spittoon was counted unhygienic.

If we can abstract pathogenicity and hygiene from our notion of dirt, we are left with the old definition of dirt as matter out of place. This is a very suggestive approach. It implies two conditions: a set of ordered relations and a contravention of that order. Dirt then, is never a unique, isolated event. Where there is dirt there is system. Dirt is the by-product of a systematic ordering and classification of matter, in so far as ordering involves rejecting inappropriate elements. This idea of dirt takes us straight into the field of symbolism and promises a link-up with more obviously symbolic systems of purity.

We can recognise in our own notions of dirt that we are using a kind of omnibus compendium which includes all the rejected elements of ordered systems. It is a relative idea. Shoes are not dirty in themselves, but it is dirty to place them on the dining-table; food is not dirty in itself, but it is dirty to leave cooking utensils in the bedroom, or food bespattered on clothing; similarly, bathroom equipment in the drawing room; clothing lying on chairs; out-door things in-doors; upstairs things downstairs; under-clothing appearing where over-clothing should be, and so on. In short, our pollution behaviour is the reaction which condemns any object or idea likely to confuse or contradict cherished classifications.

We should now force ourselves to focus on dirt. Defined in this way it appears as a residual category, rejected from our normal scheme of classifications. In trying to focus on it we run against our strongest mental habit. For it seems that whatever we perceive is organised into patterns for which we, the perceivers, are largely responsible. Perceiving is not a matter of passively allowing an organ—say of sight or hearing—to receive a ready-made impression from without, like a palette receiving a spot of paint. Recognising and remembering are not matters of stirring up old images of past impressions. It is generally agreed that all our impressions are schematically determined from the start. As perceivers we select from all the stimuli falling on our senses only those which interest us, and our interests are governed by a pattern-making tendency, sometimes called *schema* (see Bartlett, 1932). In a chaos of shifting impressions, each of us constructs a stable world in which objects have recognisable shapes, are located in depth, and have permanence. In perceiving we are building, taking some cues and rejecting others. The most acceptable cues are those which fit most easily into the pattern that is being built up. Ambiguous ones tend to be treated as if they harmonised with the rest of the pattern. Discordant ones tend to be rejected. If they are accepted the structure of assumptions has to be modified. As learning proceeds objects are named. Their names then affect the way they are perceived next time: once labelled they are more speedily slotted into the pigeon-holes in future.

As time goes on and experiences pile up, we make a greater and greater investment in our system of labels. So a conservative bias is built in. It gives us confidence. At any time we may have to modify our structure of assumptions to accommodate new experience, but the more consistent experience is with the past, the more confidence we can have in our assumptions. Uncomfortable facts which refuse to be fitted in, we find ourselves ignoring or distorting so that they do not disturb these established assumptions. By and large anything we take note of is pre-selected and organised in the very act of perceiving. We share with other animals a kind of filtering mechanism which at first only lets in sensations we know how to use.

But what about the other ones? What about the possible experiences which do not pass the filter? Is it possible to force attention into less habitual tracks? Can we even examine the filtering mechanism itself?

We can certainly force ourselves to observe things which our schematising tendencies have caused us to miss. It always gives a jar to find our first facile observation at fault. Even to gaze steadily at distorting apparatus makes some people feel physically sick, as if their own balance was attacked. Mrs. Abercrombie put a group of medical students through a course of experiments designed to show them the high degree of selection we use in the simplest observations. 'But you can't have all the world a jelly,' one protested. 'It is as though my world has been cracked open,' said another. Others reacted in a more strongly hostile way (p. 131).

But it is not always an unpleasant experience to confront ambiguity. Obviously it is more tolerable in some areas than in others. There is a whole gradient on which laughter, revulsion and shock belong at different points and intensities. The experience can be stimulating. The richness of poetry depends on the use of ambiguity, as Empson has shown. The possibility of seeing a sculpture equally well as a landscape or as a reclining nude enriches the work's interest. Ehrenzweig has even argued that we enjoy works of art because they enable us to go behind the explicit structures of our normal experience. Aesthetic pleasure arises from the perceiving of inarticulate forms.

I apologise for using anomaly and ambiguity as if they were synonymous. Strictly they are not: an anomaly is an element which does not fit a given set or series; ambiguity is a character of statements capable of two interpretations. But reflection on examples shows that there is very little advantage in distinguishing between these two terms in their practical application. Treacle is neither liquid nor solid; it could be said to give an ambiguous sense-impression. We can also say that treacle is anomalous in the classification of liquids and solids, being in neither one nor the other set.

Granted, then, that we are capable of confronting anomaly.

When something is firmly classed as anomalous the outline of the set in which it is not a member is clarified. To illustrate this I quote from Sartre's essay on stickiness. Viscosity, he says, repels in its own right, as a primary experience. An infant, plunging its hands into a jar of honey, is instantly involved in contemplating the formal properties of solids and liquids and the essential relation between the subjective experiencing self and the experiencd world (1943, p. 696 seq.). The viscous is a state half-way between solid and liquid. It is like a cross-section in a process of change. It is unstable, but it does not flow. It is soft, yielding and compressible. There is no gliding on its surface. Its stickiness is a trap, it clings like a leech; it attacks the boundary between myself and it. Long columns falling off my fingers suggest my own substance flowing into the pool of stickiness. Plunging into water gives a different impression. I remain a solid, but to touch stickiness is to risk diluting myself into viscosity. Stickiness is clinging, like a too-possessive dog or mistress. In this way the first contact with stickiness enriches a child's experience. He has learnt something about himself and the properties of matter and the interrelation between self and other things.

I cannot do justice, in shortening the passage, to the marvellous reflections to which Sartre is provoked by the idea of stickiness as an aberrant fluid or a melting solid. But it makes the point that we can and do reflect with profit on our main classifications and on experiences which do not exactly fit them. In general these reflections confirm our confidence in the main classifications. Sartre argues that melting, clinging viscosity is judged an ignoble form of existence in its very first manifestations. So from these earliest tactile adventures we have always known that life does not conform to our most simple categories.

There are several ways of treating anomalies. Negatively, we can ignore, just not perceive them, or perceiving we can condemn. Positively we can deliberately confront the anomaly and try to create a new pattern of reality in which it has a place. It is not impossible for an individual to revise his own personal scheme of classifications. But no individual lives in isolation and his scheme will have been partly received from others.

Culture, in the sense of the public, standardised values of a community, mediates the experience of individuals. It provides in advance some basic categories, a positive pattern in which ideas and values are tidily ordered. And above all, it has authority, since each is induced to assent because of the assent of others. But its public character makes its categories more rigid. A private person may revise his pattern of assumptions or not. It is a private matter. But cultural categories are public matters. They cannot so easily be subject to revision. Yet they cannot neglect the challenge of aberrant forms. Any given system of classification must give rise to anomalies, and any given culture must confront events which seem to defy its assumptions. It cannot ignore the anomalies which its scheme produces, except at risk of forfeiting confidence. This is why, I suggest, we find in any culture worthy of the name various provisions for dealing with ambiguous or anomalous events.

First, by settling for one or other interpretation, ambiguity is often reduced. For example, when a monstrous birth occurs, the defining lines between humans and animals may be threatened. If a monstrous birth can be labelled an event of a peculiar kind the categories can be restored. So the Nuer treat monstrous births as baby hippopotamuses, accidentally born to humans and, with this labelling, the appropriate action is clear. They gently lay them in the river where they belong (Evans-Pritchard, 1956, p. 84).

Second, the existence of anomaly can be physically controlled. Thus in some West African tribes the rule that twins should be killed at birth eliminates a social anomaly, if it is held that two humans could not be born from the same womb at the same time. Or take night-crowing cocks. If their necks are promptly wrung, they do not live to contradict the definition of a cock as a bird that crows at dawn.

Third, a rule of avoiding anomalous things affirms and strengthens the definitions to which they do not conform. So where Leviticus abhors crawling things, we should see the abomination as the negative side of the pattern of things approved.

Fourth, anomalous events may be labelled dangerous. Admittedly individuals sometimes feel anxiety confronted with anomaly. But it would be a mistake to treat institutions as if they evolved in the same way as a person's spontaneous reactions. Such public beliefs are more likely to be produced in the course of reducing dissonance between individual and general interpretations. Following the work of Festinger it is obvious that a person when he finds his own convictions at variance with those of friends, either wavers or tries to convince the friends of their error. Attributing danger is one way of putting a subject above dispute. It also helps to enforce conformity, as we shall show below in a chapter on morals (Chapter 8).

Fifth, ambiguous symbols can be used in ritual for the same ends as they are used in poetry and mythology, to enrich meaning or to call attention to other levels of existence. We shall see in the last chapter how ritual, by using symbols of anomaly, can incorporate evil and death along with life and goodness, into a single, grand, unifying pattern.

To conclude, if uncleanness is matter out of place, we must approach it through order. Uncleanness or dirt is that which must not be included if a pattern is to be maintained. To recognise this is the first step towards insight into pollution. It involves us in no clear-cut distinction between sacred and secular. The same principle applies throughout. Furthermore, it involves no special distinction between primitives and moderns: we are all subject to the same rules. But in the primitive culture the rule of patterning works with greater force and more total comprehensiveness. With the moderns it applies to disjointed, separate areas of existence.

3

The Abominations of Leviticus

DEFILEMENT is never an isolated event. It cannot occur except in view of a systematic ordering of ideas. Hence any piecemeal interpretation of the pollution rules of another culture is bound to fail. For the only way in which pollution ideas make sense is in reference to a total structure of thought whose key-stone, boundaries, margins and internal lines are held in relation by rituals of separation.

To illustrate this I take a hoary old puzzle from biblical scholarship, the abominations of Leviticus, and particularly the dietary rules. Why should the camel, the hare and the rock badger be unclean? Why should some locusts, but not all, be unclean? Why should the frog be clean and the mouse and the hippopotamus unclean? What have chameleons, moles and crocodiles got in common that they should be listed together (Levi. xi, 27)?

To help follow the argument I first quote the relevant versions of Leviticus and Deuteronomy using the text of the New Revised Standard Translation.

Deut. xiv

3. You shall not eat any abominable things. 4. These are the animals you may eat: the ox, the sheep, the goat, 5. the hart, the gazelle, the roe-buck, the wild goat, the ibex, the antelope and the mountain-sheep. 6. Every animal that parts the hoof and has the hoof cloven in two, and chews the cud, among the animals you may eat. 7. Yet of those that chew the cud or have the hoof cloven you shall not eat these: The camel, the hare and the rock badger, because they chew the cud but do not part the hoof, are unclean for you. 8. And the swine, because it parts the hoof

41

but does not chew the cud, is unclean for you. Their flesh you shall not eat, and their carcasses you shall not touch. 9. Of all that are in the waters you may eat these: whatever has fins and scales you may eat. 10. And whatever does not have fins and scales you shall not eat; it is unclean for you. 11. You may eat all clean birds. 12. But these are the ones which you shall not eat: the eagle, the vulture, the osprey. 13. the buzzard, the kite, after their kinds; 14. every raven after its kind; 15. the ostrich, the night hawk, the sea gull, the hawk, after their kinds; 16. the little owl and the great owl, the water hen 17. and the pelican, the carrion vulture and the cormorant, 18. the stork, the heron, after their kinds; the hoopoe and the bat. 19. And all winged insects are unclean for you; they shall not be eaten. 20. All clean winged things you may eat.

Lev. xi

2. These are the living things which you may eat among all the beasts that are on the earth. 3. Whatever parts the hoof and is cloven-footed and chews the cud, among the animals you may eat. 4. Nevertheless among those that chew the cud or part the hoof, you shall not eat these: The camel, because it chews the cud but does not part the hoof, is unclean to you. 5. And the rock badger, because it chews the cud but does not part the hoof, is unclean to you. 6. And the hare, because it chews the cud but does not part the hoof, is unclean to you. 7. And the swine, because it parts the hoof and is cloven-footed but does not chew the cud, is unclean to you. 8. Of their flesh you shall not eat, and their carcasses you shall not touch; they are unclean to you. 9. These you may eat of all that are in the waters. Everything in the waters that has fins and scales, whether in the seas or in the rivers, you may eat. 10. But anything in the seas or the rivers that has not fins and scales, of the swarming creatures in the waters and of the living creatures that are in the waters, is an abomination to you. 11. They shall remain an abomination to you; of their flesh you shall not eat, and their carcasses you shall have in abomination. 12. Everything in the waters that has not fins and scales is an abomination to you. 13. And these you shall have in abomination among the birds, they shall not be eaten, they are an abomination: the eagle, the ossifrage, the osprey, 14. the kite, the falcon according to its kind, 15. every raven according to its kind, 16. the

ostrich and the night hawk, the sea gull, the hawk according to its kind, 17. the owl, the cormorant, the ibis, 18. the water hen, the pelican, the vulture, 19. the stork, the heron according to its kind, the hoopoe and the bat. 20. All winged insects that go upon all fours are an abomination to you. 21. Yet among the winged insects that go on all fours you may eat those which have legs above their feet, with which to leap upon the earth. 22. Of them you may eat: the locust according to its kind, the bald locust according to its kind, the cricket according to its kind, and the grasshopper according to its kind. 23. But all other winged insects which have four feet are an abomination to you. 24. And by these you shall become unclean; whoever touches their carcass shall be unclean until the evening, 25. and whoever carries any part of their carcass shall wash his clothes and be unclean until the evening. 26. Every animal which parts the hoof but is not cloven-footed or does not chew the cud is unclean to you: everyone who touches them shall be

unclean. 27. And all that go on their paws, among the animals that go on all fours, are unclean to you; whoever touches their carcass shall be unclean until the evening, 28. and he who carries their carcass shall wash his clothes and be unclean until the evening; they are unclean to you. 29. And these are unclean to you among the swarming things that swarm upon the earth; the weasel, the mouse, the great lizard according to its kind, 30. the gecko, the land crocodile, the lizard, the sand lizard and the chameleon. 31. These are unclean to you among all that swarm; whoever touches them when they are dead shall be unclean until the evening. 32. And anything upon which any of them falls when they are dead shall be unclean.

41. Every swarming thing that swarms upon the earth is an abomination; it shall not be eaten. 42. Whatever goes on its belly, and whatever goes on all fours, or whatever has many feet, all the swarming things that swarm upon the earth, you shall not eat; for they are an abomination.

All the interpretations given so far fall into one of two groups: either the rules are meaningless, arbitrary because their intent is disciplinary and not doctrinal, or they are allegories of virtues and vices. Adopting the view that religious prescriptions are largely devoid of symbolism, Maimonides said:

'The Law that sacrifices should be brought is evidently of great use . . . but we cannot say why one offering should be a lamb whilst another is a ram, and why a fixed number of these should be brought. Those who trouble themselves to find a cause for any of these detailed rules are in my eyes devoid of sense . . . .'

As a mediaeval doctor of medicine, Maimonides was also disposed to believe that the dietary rules had a sound physiological basis, but we have already dismissed in the second chapter the medical approach to symbolism. For a modern version of the view that the dietary rules are not symbolic, but ethical, disciplinary, see Epstein's English notes to the Babylonian Talmud and also his popular history of Judaism (1959, p. 24):

'Both sets of laws have one common aim . . . Holiness. While the positive precepts have been ordained for the cultivation of virtue, and for the promotion of those finer qualities which distinguish the truly religious and ethical being, the negative precepts are defined to combat vice and suppress other evil tendencies and instincts which stand athwart man's striving towards holiness. . . . The negative religious laws are likewise assigned educational aims and purposes. Foremost among these is the prohibition of eating the flesh of certain animals classed as 'unclean'. This law has nothing totemic about it. It is expressly associated in Scripture with the ideal of holiness. Its real object is to train the Israelite in self-control as the indispensable first step for the attainment of holiness.'

According to Professor Stein's *The Dietary Laws in Rabbinic and Patristic Literature*, the ethical interpretation goes back to the time of Alexander the Great and the Hellenic influence on Jewish culture. The first century A.D. letters of Aristeas teaches that not only are the Mosaic rules a valuable discipline which 'prevents the Jews from thoughtless action and injustice', but they also coincide with what natural reason would dictate for achieving the good life. So the Hellenic influence allows the medical and ethical interpretations to run together. Philo held that Moses' principle of selection was precisely to choose the most delicious meats:

'The lawgiver sternly forbade all animals of land, sea or air whose flesh is the finest and fattest, like that of pigs and scaleless fish, knowing that they set a trap for the most slavish of senses, the taste, and that they produced gluttony',

(and here we are led straight into the medical interpretation)

> 'an evil dangerous to both soul and body, for gluttony begets indigestion, which is the source of all illnesses and infirmities'.

In another stream of interpretation, following the tradition of Robertson Smith and Frazer, the Anglo-Saxon Old Testament scholars have tended to say simply that the rules are arbitrary because they are irrational. For example, Nathaniel Micklem says :

> 'Commentators used to give much space to a discussion of the question why such and such creatures, and such or such states and symptoms were unclean. Have we, for instance, primitive rules of hygiene? Or were certain creatures and states unclean because they represented or typified certain sins? It may be taken as certain that neither hygiene, nor any kind of typology, is the basis of uncleanness. These regulations are not by any means to be rationalised. Their origins may be diverse, and go back beyond history . . .'

Compare also R. S. Driver (1895):

> 'The principle, however, determining the line of demarcation between clean animals and unclean, is not stated; and what it is has been much debated. No single principle, embracing all the cases, seems yet to have been found, and not improbably more principles than one co-operated. Some animals may have been prohibited on account of their repulsive appearance or uncleanly habits, others upon sanitary grounds; in other cases, again, the motive of the prohibition may very probably have been a religious one, particularly animals may have been supposed, like the serpent in Arabia, to be animated by super-human or demoniac beings, or they may have had a sacramental significance in the heathen rites of other nations; and the prohibition may have been intended as a protest against these beliefs. . . .'

P. P. Saydon takes the same line in the *Catholic Commentary on Holy Scripture* (1953), acknowledging his debt to Driver and to Robertson Smith. It would seem that when Robertson Smith applied the ideas of primitive, irrational and unexplainable to some parts of Hebrew religion they remained thus labelled and unexamined to this day.

Needless to say such interpretations are not interpretations at all, since they deny any significance to the rules. They express bafflement in a learned way. Micklem says it more frankly when he says of Leviticus:

> 'Chapters xi to xv are perhaps the least attractive in the whole Bible. To the modern reader there is much in them that is meaningless or repulsive. They are concerned with ritual 'uncleanness' in respect of animals (11) of childbirth (12), skin diseases and stained garments (13), of the rites for the purgation of skin diseases (14) of leprosy and of various issues or secretions of the human body (15). Of what interest can such subjects be except to the anthropologist? What can all this have to do with religion?'

Pfeiffer's general position is to be critical of the priestly and legal elements in the life of Israel. So he too lends his authority to the view that the rules in the Priestly Code are largely arbitrary:

> 'Only priests who were lawyers could have conceived of religion as a theocracy regulated by a divine law fixing exactly, and therefore arbitrarily, the sacred obligations of the people to their God. They thus sanctified the external, obliterated from religion both the ethical ideals of Amos and the tender emotions of Hosea, and reduced the Universal Creator to the stature of an inflexible despot. . . . From immemorial custom P derived the two fundamental notions which characterise its legislation: physical holiness and arbitrary enactment—archaic conceptions which the reforming prophets had discarded in favour of spiritual holiness and moral law.' (p. 91)

It may be true that lawyers tend to think in precise and codified forms. But is it plausible to argue that they tend to codify sheer nonsense—arbitrary enactments? Pfeiffer tries to have it both ways, insisting on the legalistic rigidity of the priestly authors and pointing to the lack of order in the setting out of the chapter to justify his view that the rules are arbitrary. Arbitrariness is a decidedly unexpected quality to find in Leviticus, as the Rev. Prof. H. J. Richards has pointed out to me. For source criticism attributes Leviticus to the Priestly source, the dominant concern of whose authors was for order. So the weight of source criticism supports us in looking for another interpretation.

As for the idea that the rules are allegories of virtues and vices, Professor Stein derives this vigorous tradition from the same early Alexandrian influence on Jewish thought (p. 145 seq.). Quoting the letter of Aristeas, he says that the High Priest, Eleazar:

'admits that most people find the biblical food restrictions not understandable. If God is the Creator of everything, why should His law be so severe as to exclude some animals even from touch (128 f)? His first answer still links the dietary restrictions with the danger of idolatry. . . . The second answer attempts to refute specific charges by means of allegorical exegesis. Each law about forbidden foods has its deep reason. Moses did not enumerate the mouse or the weasel out of a special consideration for them (143 f). On the contrary, mice are particularly obnoxious because of their destructiveness, and weasels, the very symbol of malicious tale-bearing, conceive through the ear and give birth through the mouth (164 f). Rather have these holy laws been given for the sake of justice to awaken in us devout thoughts and to form our character (161-168). The birds, for instance, the Jews are allowed to eat are all tame and clean, as they live by corn only. Not so the wild and carnivorous birds who fall upon lambs and goats, and even human beings. Moses, by calling the latter unclean, admonished the faithful not to do violence to the weak and not to trust their own power (145-148). Cloven-hoofed animals which part their hooves symbolise that all our actions must betray proper ethical distinction and be directed towards righteousness. . . . Chewing the cud, on the other hand stands for memory.'

Professor Stein goes on to quote Philo's use of allegory to interpret the dietary rules:

'Fish with fins and scales, admitted by the law, symbolise endurance and self-control, whilst the forbidden ones are swept away by the current, unable to resist the force of the stream. Reptiles, wriggling along by trailing their belly, signify persons who devote themselves to their ever greedy desires and passions. Creeping things, however, which have legs above their feet, so that they can leap, are clean because they symbolise the success of moral efforts.'

Christian teaching has readily followed the allegorising tradition. The first century epistle of Barnabus, written to convince the Jews that their law had found its fulfilment, took the clean and unclean animals to refer to various types of men, leprosy to mean sin, etc. A more recent example of this tradition is in

Bishop Challoner's notes on the Westminster Bible in the beginning of this century:

> 'Hoof divided and cheweth the cud. The dividing of the hoof and chewing of the cud signify discretion between good and evil, and meditating on the law of God; and where either of these is wanting, man is unclean. In like manner fishes were reputed unclean that had not fins and scales: that is souls that did not raise themselves up by prayer and cover themselves with the scales of virtue.' Footnote verse 3.

These are not so much interpretations as pious commentaries. They fail as interpretations because they are neither consistent nor comprehensive. A different explanation has to be developed for each animal and there is no end to the number of possible explanations.

Another traditional approach, also dating back to the letter of Aristeas, is the view that what is forbidden to the Israelites is forbidden solely to protect them from foreign influence. For instance, Maimonides held that they were forbidden to seethe the kid in the milk of its dam because this was a cultic act in the religion of the Canaanites. This argument cannot be comprehensive, for it is not held that the Israelites consistently rejected all the elements of foreign religions and invented something entirely original for themselves. Maimonides accepted the view that some of the more mysterious commands of the law had as their object to make a sharp break with heathen practices. Thus the Israelites were forbidden to wear garments woven of linen and wool, to plant different trees together, to have sexual intercourse with animals, to cook meat with milk, simply because these acts figured in the rites of their heathen neighbours. So far, so good: the laws were enacted as barriers to the spread of heathen styles of ritual. But in that case why were some heathen practices allowed? And not only allowed—if sacrifice be taken as a practice common to heathens and Israelites—but given an absolutely central place in the religion. Maimonides' answer, at any rate in *The Guide to the Perplexed*, was to justify sacrifice as a transitional stage, regrettably heathen, but necessarily allowed because it would be impractical to wean the Israelites abruptly from their heathen past. This is an extraordinary statement to come from the pen of a rabbinical scholar, and indeed in his serious rabbinical writings Maimonides did not

The Abominations of Leviticus

attempt to maintain the argument: on the contrary, he there counted sacrifice as the most important act of the Jewish religion.

At least Maimonides saw the inconsistency and was led by it into contradiction. But later scholars seem content to use the foreign influence argument one way or the other, according to the mood of the moment. Professor Hooke and his colleagues have clearly established that the Israelites took over some Canaanite styles of worship, and the Canaanites obviously had much in common with Mesopotamian culture (1933). But it is no explanation to represent Israel as a sponge at one moment and as a repellent the next, without explaining why it soaked up this foreign element but repelled that one. What is the value of saying that seething kids in milk and copulating with cows are forbidden in Leviticus because they are the fertility rites of foreign neighbours (1935), since Israelites took over other foreign rites? We are still perplexed to know when the sponge is the right or the wrong metaphor. The same argument is equally puzzling in Eichrodt (pp. 230-1). Of course no culture is created out of nothing. The Israelites absorbed freely from their neighbours, but not quite freely. Some elements of foreign culture were incompatible with the principles of patterning on which they were constructing their universe; others were compatible. For instance, Zaehner suggests that the Jewish abomination of creeping things may have been taken over from Zoroastrianism (p. 162). Whatever the historical evidence for this adoption of a foreign element into Judaism, we shall see that there was in the patterning of their culture a pre-formed compatibility between this particular abomination and the general principles on which their universe was constructed.

Any interpretations will fail which take the Do-nots of the Old Testament in piecemeal fashion. The only sound approach is to forget hygiene, aesthetics, morals and instinctive revulsion, even to forget the Canaanites and the Zoroastrian Magi, and start with the texts. Since each of the injunctions is prefaced by the command to be holy, so they must be explained by that command. There must be contrariness between holiness and abomination which will make over-all sense of all the particular restrictions.

Holiness is the attribute of Godhead. Its root means 'set apart'. What else does it mean? We should start any cosmological enquiry by seeking the principles of power and danger. In the Old

Testament we find blessing as the source of all good things, and the withdrawal of blessing as the source of all dangers. The blessing of God makes the land possible for men to live in.

God's work through the blessing is essentially to create order, through which men's affairs prosper. Fertility of women, live-stock and fields is promised as a result of the blessing and this is to be obtained by keeping covenant with God and observing all His precepts and ceremonies (Deut. XXVIII, 1-14). Where the blessing is withdrawn and the power of the curse unleashed, there is barrenness, pestilence, confusion. For Moses said:

'But if you will not obey the voice of the Lord your God or be careful to do all his commandments and his statutes which I command you to this day, then all these curses shall come upon you and overtake you. Cursed shall you be in the city, and cursed shall you be in the field. Cursed shall be your basket and your kneading trough. Cursed shall be the fruit of your body, and the fruit of your ground, the increase of your cattle, and the young of your flock. Cursed shall you be when you come in and cursed shall you be when you go out. The Lord will send upon you curses, confusion, and frustration in all that you undertake to do, until you are destroyed and perish quickly on account of the evil of your doings, because you have for-saken me . . . The Lord will smite you with consumption, and with fever, inflammation, and fiery heat, and with drought, and with blasting and with mildew; they shall pursue you till you perish. And the heavens over your head shall be brass and the earth under you shall be iron. The Lord will make the rain of your land powder and dust; from heaven it shall come down upon you until you are destroyed.' (Deut. XXVIII, 15-24)

From this it is clear that the positive and negative precepts are held to be efficacious and not merely expressive: observing them draws down prosperity, infringing them brings danger. We are thus entitled to treat them in the same way as we treat primitive ritual avoidances whose breach unleashes danger to men. The precepts and ceremonies alike are focussed on the idea of the holiness of God which men must create in their own lives. So this is a universe in which men prosper by conforming to holiness and perish when they deviate from it. If there were no other clues we should be able to find out the Hebrew idea of the holy by examining the precepts by which men conform to it. It is evidently not goodness in the sense of an all-embracing humane kindness. Justice and moral goodness may well illustrate holiness and form part of it, but holiness embraces other ideas as well.

Granted that its root means separateness, the next idea that emerges is of the Holy as wholeness and completeness. Much of Leviticus is taken up with stating the physical perfection that is required of things presented in the temple and of persons approaching it. The animals offered in sacrifice must be without blemish, women must be purified after childbirth, lepers should be separated and ritually cleansed before being allowed to approach it once they are cured. All bodily discharges are defiling and disqualify from approach to the temple. Priests may only come into contact with death when their own close kin die. But the high priest must never have contact with death.

Levit. xxi

17. 'Say to Aaron, None of your descendants throughout their generations who has a blemish may approach to offer the bread of his God. 18. For no one who has a blemish shall draw near, a man blind or lame, or one who has a mutilated face or a limb too long. 19. or a man who has an injured foot or an injured hand, 20. or a hunch-back, or a dwarf, or a man with a defect in his sight or an itching disease or scabs, or crushed testicles; 21. no man of the descendants of Aaron the priest who has a blemish shall come near to offer the Lord's offerings by fire; . . .'

In other words, he must be perfect as a man, if he is to be a priest.

This much reiterated idea of physical completeness is also worked out in the social sphere and particularly in the warriors' camp. The culture of the Israelites was brought to the pitch of greatest intensity when they prayed and when they fought. The army could not win without the blessing and to keep the blessing in the camp they had to be specially holy. So the camp was to be preserved from defilement like the Temple. Here again all bodily discharges disqualified a man from entering the camp as they would disqualify a worshipper from approaching the altar. A warrior who had had an issue of the body in the night should keep outside the camp all day and only return after sunset, having washed. Natural functions producing bodily waste were to be performed outside the camp (Deut. XXIII, 10-15). In short the idea of holiness was given an external, physical expression in the wholeness of the body seen as a perfect container.

Wholeness is also extended to signify completeness in a social context. An important enterprise, once begun, must not be left incomplete. This way of lacking wholeness also disqualifies a man from fighting. Before a battle the captains shall proclaim:

Deut. xx

5. 'What man is there that has built a new house and has not dedicated it? Let him go back to his house, lest he die in the battle and another man dedicate it. 6. What man is there that has planted a vineyard and has not enjoyed its fruit? Let him go back to his house, lest he die in the battle and another man enjoy its fruit. 7. And what man is there that hath betrothed a wife and has not taken her? Let him go back to his house, lest he die in the battle and another man take her.'

Admittedly there is no suggestion that this rule implies defilement. It is not said that a man with a half-finished project on his hands is defiled in the same way that a leper is defiled. The next verse in fact goes on to say that fearful and faint-hearted men should go home lest they spread their fears. But there is a strong suggestion in other passages that a man should not put his hand to the plough and then turn back. Pedersen goes so far as to say that:

'in all these cases a man has started a new important under-taking without having finished it yet . . . a new totality has come into existence. To make a breach in this prematurely, i.e. before it has attained maturity or has been finished, in-volves a serious risk of sin'. (Vol. III, p. 9)

If we follow Pedersen, then blessing and success in war required a man to be whole in body, whole-hearted and trailing no uncompleted schemes. There is an echo of this actual passage in the New Testament parable of the man who gave a great feast and whose invited guests incurred his anger by making excuses (Luke xiv, 16-24; Matt. xxii. See Black & Rowley, 1962, p. 836). One of the guests had bought a new farm, one had bought ten oxen and had not yet tried them, and one had married a wife. If according to the old Law each could have validly justified his refusal by reference to Deut. xx, the parable supports Pedersen's view that interruption of new projects was held to be bad in civil as well as military contexts.

Other precepts develop the idea of wholeness in another direction. The metaphors of the physical body and of the new undertaking relate to the perfection and completeness of the individual and his work. Other precepts extend holiness to species and categories. Hybrids and other confusions are abominated.

Lev. xviii

'23. And you shall not lie with any beast and defile yourself with it, neither shall any woman give herself to a beast to lie with it: it is perversion.'

The word 'perversion' is a significant mistranslation of the rare Hebrew word *tebhel*, which has as its meaning mixing or confusion. The same theme is taken up in Leviticus xix, 19.

'You shall keep my statutes. You shall not let your cattle breed with a different kind; you shall not sow your field with two kinds of seed; nor shall there come upon you a garment of cloth made of two kinds of stuff.'

All these injunctions are prefaced by the general command:

'Be holy, for I am holy.'

We can conclude that holiness is exemplified by completeness. Holiness requires that individuals shall conform to the class to which they belong. And holiness requires that different classes of things shall not be confused.

Another set of precepts refines on this last point. Holiness means keeping distinct the categories of creation. It therefore involves correct definition, discrimination and order. Under this head all the rules of sexual morality exemplify the holy. Incest and adultery (Lev. xviii, 6-20) are against holiness, in the simple sense of right order. Morality does not conflict with holiness, but holiness is more a matter of separating that which should be separated than of protecting the rights of husbands and brothers.

Then follows in chapter xix another list of actions which are contrary to holiness. Developing the idea of holiness as order, not confusion, this list upholds rectitude and straight-dealing as holy, and contradiction and double-dealing as against holiness. Theft, lying, false witness, cheating in weights and measures, all kinds of dissembling such as speaking ill of the deaf (and presumably smiling to their face), hating your brother in your heart (while presumably speaking kindly to him), these are clearly contradictions between what seems and what is. This chapter also says much about generosity and love, but these are positive commands, while I am concerned with negative rules.

We have now laid a good basis for approaching the laws about clean and unclean meats. To be holy is to be whole, to be one; holiness is unity, integrity, perfection of the individual and of the kind. The dietary rules merely develop the metaphor of holiness on the same lines.

First we should start with livestock, the herds of cattle, camels, sheep and goats which were the livelihood of the Israelites. These animals were clean inasmuch as contact with them did not require purification before approaching the Temple. Livestock, like the inhabited land, received the blessing of God. Both land and livestock were fertile by the blessing, both were drawn into the divine order. The farmer's duty was to preserve the blessing. For one thing, he had to preserve the order of creation. So no hybrids, as we have seen, either in the fields or in the herds or in the clothes made from wool or flax. To some extent men covenanted with their land and cattle in the same way as God covenanted with them. Men respected the first born of their cattle, obliged them to keep the Sabbath. Cattle were literally domesticated as slaves. They had to be brought into the social order in order to enjoy the blessing. The difference between cattle and the wild beasts is that the wild beasts have no covenant to protect them. It is possible that the Israelites were like other pastoralists who do not relish wild game. The Nuer of the South Sudan, for instance, apply a sanction of disapproval of a man who lives by hunting. To be driven to eating wild meat is the sign of a poor herdsman. So it would be probably wrong to think of the Israelites as longing for forbidden meats and finding the restrictions irksome. Driver is surely right in taking the rules as an *a posteriori* generalisation of their habits. Cloven-hoofed, cud-chewing ungulates are the model of the proper kind of food for a pastoralist. If they must eat wild game, they can eat wild game that shares these distinctive characters and is therefore of the same general species. This is a kind of casuistry which permits scope for hunting antelope and wild goats and wild sheep. Everything would be quite straightforward were it not that the legal mind has seen fit to give ruling on some border-line cases. Some animals seem to be ruminant, such as the hare and the hyrax (or rock badger), whose constant grinding of their teeth was held to be cud-chewing. But they are definitely not cloven-hoofed and so are excluded by name. Similarly for animals which are cloven-hoofed but are not ruminant, the pig and the camel. Note that this failure to conform to the two necessary criteria for defining cattle is the only reason given in the Old Testament for avoiding the pig; nothing whatever is said about its dirty scavenging habits. As the pig does not yield milk, hide nor wool, there is no other reason for keeping it except for its flesh. And if the Israelites did not keep pig they would not be familiar with its habits. I suggest that originally the sole reason for its being counted as unclean is its failure as a wild boar to get into the antelope class, and that in this it is on the same footing as the camel and the hyrax, exactly as is stated in the book.

After these borderline cases have been dismissed, the law goes on to deal with creatures according to how they live in the three elements, the water, the air and the earth. The principles here applied are rather different from those covering the camel, the pig, the hare and the hyrax. For the latter are excepted from clean food in having one but not both of the defining characters of livestock. Birds I can say nothing about, because, as I have said, they are named and not described and the translation of the name is open to doubt. But in general the underlying principle of cleanness in animals is that they shall conform fully to their class. Those species are unclean which are imperfect members of their class, or whose class itself confounds the general scheme of the world.

To grasp this scheme we need to go back to Genesis and the creation. Here a three-fold classification unfolds, divided between the earth, the waters and the firmament. Leviticus takes up this scheme and allots to each element its proper kind of animal life. In the firmament two-legged fowls fly with wings. In the water scaly fish swim with fins. On the earth four-legged animals hop, jump or walk. Any class of creatures which is not equipped for the right kind of locomotion in its element is contrary to holiness. Contact with it disqualifies a person from approaching the Temple. Thus anything in the water which has not fins and scales is unclean (XI, 10-12). Nothing is said about predatory habits or of scavenging. The only sure test for cleanness in a fish is its scales and its propulsion by means of fins.

Four-footed creatures which fly (XI, 20-26) are unclean. Any creature which has two legs and two hands and which goes on all fours like a quadruped is unclean (xi, 27). Then follows (v. 29) a much disputed list. On some translations, it would appear to consist precisely of creatures endowed with hands instead of front feet, which perversely use their hands for walking: the weasel, the mouse, the crocodile, the shrew, various kinds of lizards, the chameleon and mole (Danby, 1933), whose forefeet are uncannily hand-like. This feature of this list is lost in the New Revised Standard Translation which uses the word 'paws' instead of hands.

The last kind of unclean animal is that which creeps, crawls or swarms upon the earth. This form of movement is explicitly contrary to holiness (Levit. xi, 41-44). Driver and White use 'swarming' to translate the Hebrew *shérec*, which is applied to both those which teem in the waters and those which swarm on the ground. Whether we call it teeming, trailing, creeping, crawling or swarming, it is an indeterminate form of movement. Since the main animal categories are defined by their typical movement, 'swarming' which is not a mode of propulsion proper to any particular element, cuts across the basic classification. Swarming things are neither fish, flesh nor fowl. Eels and worms inhabit water, though not as fish; reptiles go on dry land, though not as quadrupeds; some insects fly, though not as birds. There is no order in them. Recall what the Prophecy of Habakkuk says about this form of life:

> 'For thou makest men like the fish of the sea, like crawling things that have no ruler.' (i, v. 14)

The prototype and model of the swarming things is the worm. As fish belong in the sea so worms belong in the realm of the grave, with death and chaos.

The case of the locusts is interesting and consistent. The test of whether it is a clean and therefore edible kind is how it moves on the earth. If it crawls it is unclean. If it hops it is clean (xi, v. 21). In the Mishnah it is noted that a frog is not listed with creeping things and conveys no uncleanness (Danby, p. 722). I suggest that the frog's hop accounts for it not being listed. If penguins lived in the Near East I would expect them to be ruled unclean as wingless birds. If the list of unclean birds could be retranslated from this point of view, it might well turn out that they are anomalous because they swim and dive as well as they fly, or in some other way they are not fully bird-like.

Surely now it would be difficult to maintain that 'Be ye Holy' means no more than 'Be ye separate'. Moses wanted the children of Israel to keep the commands of God constantly before their minds:

Deut. XI

'18. You shall therefore lay up these words of mine in your heart and in your soul; and you shall bind them as a sign upon your hand, and they shall be as frontlets between your eyes. 19. And you shall teach them to your children, talking of them when you are sitting in your house, and when you are walking by the way, and when you lie down and when you rise. 20. And you shall write them upon the doorposts of your house and upon your gates.'

If the proposed interpretation of the forbidden animals is correct, the dietary laws would have been like signs which at every turn inspired meditation on the oneness, purity and completeness of God. By rules of avoidance holiness was given a physical expression in every encounter with the animal kingdom and at every meal. Observance of the dietary rules would thus have been a meaningful part of the great liturgical act of recognition and worship which culminated in the sacrifice in the Temple.

4

Magic and Miracle

ONCE when a band of !Kung Bushmen had performed their rain rituals, a small cloud appeared on the horizon, grew and darkened. Then rain fell. But the anthropologists who asked if the Bushmen reckoned the rite had produced the rain, were laughed out of court (Marshall, 1957). How naïve can we get about the beliefs of others? Old anthropological sources are full of the notion that primitive people expect rites to produce an immediate intervention in their affairs, and they poke kindly fun at those who supplement their rituals of healing with European medicine, as if this testified to lack of faith. The Dinka perform an annual ceremony to cure malaria. The ceremony is timed for the month in which it is to be expected that malaria will soon abate. A European observer who witnessed it remarked dryly that the officiant ended by urging everyone to attend the clinic regularly if they hoped to get well (Lienhardt 1961).

It is not difficult to trace the idea that primitives expect their rites to have external efficacy. There is a comfortable assumption in the roots of our culture that foreigners know no true spiritual religion. On this assumption Frazer's grandiose description of primitive magic took root and flourished. Magic was carefully separated from other ceremonial, as if primitive tribes were populations of Ali Babas and Aladdins, uttering their magic words and rubbing their magic lamps. The European belief in primitive magic has led to a false distinction between primitive and modern cultures, and sadly inhibited comparative religion. I do not propose to show how the term magic has been used by various scholars hitherto. Too much erudition has been expended already on defining and naming symbolic actions which are held to be efficacious for altering the course of events (Goody, Gluckman).

On the continent magic has remained a vague literary term, described but never rigorously defined. It is clear that in the tradition of Mauss' *Théorie de la Magie*, the word does not connote a particular class of rituals, but rather the whole corpus of ritual and belief of primitive peoples. No special focus is centred on efficacy. We owe to Frazer the isolating and hardening of the idea of magic as the efficacious symbol (see Chapter 1). Malinowski further developed the idea uncritically and gave its currency renewed life. For Malinowski magic takes its origin in the expression of an individual's emotions. Passion, as it contorted his face, and caused the magician to stamp his foot or shake his fist, also led him to enact his strong desire for gain or revenge. This physical enactment, almost involuntary at the start, a deluded wish-fulfilment, was for him the basis of the magic rite (see Nadel, p. 194). Malinowski had such original insights into the creative effect of ordinary speech that he profoundly influenced contemporary linguistics. How could he have barrenly isolated magic rite from other rites and discussed magic as a kind of poor man's whisky, used for gaining conviviality and courage against daunting odds? This is another aberration which we can lay to the door of Frazer, whose disciple he claimed to be.

Since Robertson Smith drew a parallel between Roman Catholic ritual and primitive magic, let us gratefully take the hint. For magic let us read miracle and reflect on the relation between ritual and miracle in the minds of the mass of believers in the miracle-believing ages of Christianity. There we find that the possibility of miracle was always present; it did not necessarily depend on rite, it could be expected to erupt anywhere at any time in response to virtuous need or the demands of justice. It inhered more potently in some material objects, places and persons. It could not be laid under automatic control; the saying of the right words or sprinkling of holy water could not guarantee a cure. The power of miraculous intervention was believed to exist, but there was no certain way of harnessing it. It was as different and as like Islamic *Baraka* or Teutonic Luck or Polynesian *Mana* as each is different from the other. Each primitive universe hopes to harness some such marvellous power to the needs of men, and each supposes that a different set of links has to be reckoned with, as we shall see in the next chapter. In the miraculous period of our Christian heritage miracle did not only occur through enacted rites, nor were rites always performed in the expectation of miracle. It is realistic to suppose an equally loose relation holds between rite and magic effect in primitive religion. We should recognise that the possibility of magic intervention is always present in the mind of believers, that it is human and natural to hope for material benefits from the enactment of cosmic symbols. But it is wrong to treat primitive ritual as primarily concerned with producing magical effects. The priest in a primitive culture is not necessarily a magic wonder worker. This idea has barred our understanding of alien religions, but it is only a recent by-product of a more deep-rooted prejudice.

A contrast between interior will and exterior enactment goes deep into the history of Judaism and Christianity. Of its very nature any religion must swing between these two poles. There must be a move from internal to external religious life, if a new religion endure even a decade after its first revolutionary fervour. And finally, the hardening of the external crust becomes a scandal and provokes new revolutions.

So the rage of the Old Testament prophets was continually renewed against empty external forms paraded instead of humble and contrite hearts. From the time of the first Council of Jerusalem, the Apostles tried to take their stand on a spiritual interpretation of sanctity. The Sermon on the Mount was seen as the deliberate Messianic counterpart of the Mosaic law. St. Paul's frequent references to the law as part of the old dispensation, a bondage and a yoke, are too familiar to need quotation. From this time on the physiological condition of a person, whether leprous, bleeding, or crippled, should have become irrelevant to their capacity to approach the altar. The foods they ate, the things they touched, the days on which they did things, such accidental conditions should have no effect on their spiritual status. Sin was to be regarded as a matter of the will and not of external circumstance. But continually the spiritual intentions of the early Church were frustrated by spontaneous resistance to the idea that bodily states were irrelevant to ritual. The idea of pollution by blood, for example, seems to have been a long time dying, if we judge by some early Penitentials.

See the Penitential of Archbishop Theodore of Canterbury, A.D. 668-690:

> 'If without knowing it one eats what is polluted by blood or any unclean thing, it is nothing; but if he knows, he shall do penance according to the degree of pollution. . . .'

He also requires from women 40 days of purgation after the birth of a child, and enjoins penance of three weeks' fast on any woman, lay or religious, who enters a church, or communicates during menstruation (McNeill & Gamer).

Needless to say, these rules were not adopted as part of the Corpus of Canon Law, and now it is difficult to find instances of ritual uncleanness in Christian practice. Injunctions, which in their origin may have been concerned with removing pollution of blood, are interpreted as carrying only a symbolic spiritual significance. For example, it is usual to reconsecrate a church if blood has been shed in its precincts, but St. Thomas Aquinas explains that 'bloodshed' refers to voluntary injury leading to bloodshed, which implies sin, and that it is sin in a holy place which desecrates it, not defilement by bloodshed. Similarly, the rite for purification of a mother probably does derive ultimately from Judaic practice, but the modern Roman Ritual, which dates back to Pope Paul V (1605-21), presents the churching of women simply as an act of thanksgiving.

The long history of protestantism witnesses to the need for continual watch on the tendency of ritual form to harden and replace religious feeling. In wave upon wave the Reformation has continued to thunder against the empty encrustation of ritual. So long as Christianity has any life, it will never be time to stop echoing the parable of the Pharisee and the Publican, to stop saying that external forms can become empty and mock the truths they stand for. With every new century we become heirs to a longer and more vigorous anti-ritualist tradition.

This is right and good as far as our own religious life is concerned, but let us beware of importing uncritically a dread of dead formality in ourselves into our judgments of other religions. The Evangelical movement has left us with a tendency to suppose that any ritual is empty form, that any codifying of conduct is alien to natural movements of sympathy, and that any external religion betrays true interior religion. From this it is a short step to assuming something about primitive religions. If they are formal enough to be reported at all, they are too formal, and without interior religion. For example, Pfeiffer's *Books of the Old Testament* has this anti-ritualist basis which leads him to contrast 'the old religion of cult' with the prophets' 'new one of conduct'. He writes as if there could be no spiritual content in the old cult (pp. 55 seq.). The religious history of Israel he presents as if the stern, insensitive lawgivers were in conflict with the prophets, and never allows that both could have been engaged in the same service, or that ritual and codification could have something to do with spirituality. According to Pfeiffer the lawyer priests:

> 'sanctified the external, obliterated from religion the ethical ideals of Amos and the tender emotions of Hosea, and reduced the universal creator to the status of an inflexible despot. . . . From immemorial custom P derived the two fundamental notions which characterised its legislation: physical holiness and arbitrary enactment—archaic conceptions which the re-forming prophets had discarded in favour of spiritual holiness and moral law.' (p. 91)

This is not history, but sheer anti-ritualist prejudice. For it is a mistake to suppose that there can be religion which is all interior, with no rules, no liturgy, no external signs of inward states. As with society, so with religion, external form is the condition of its existence. As the heirs of the Evangelical tradition we have been brought up to suspect formality and to look for spontaneous expressions like the Minister's sister whom Mary Webb made to say, 'Home-made cakes and home-made prayers are always best'. As a social animal, man is a ritual animal. If ritual is suppressed in one form it crops up in others, more strongly the more intense the social interaction. Without the letters of condolence, telegrams of congratulations and even occasional postcards, the friendship of a separated friend is not a social reality. It has no existence without the rites of friendship. Social rituals create a reality which would be nothing without them. It is not too much to say that ritual is more to society than words are to thought. For it is very possible to know something and then find words for it. But it is impossible to have social relations without symbolic acts.

We shall understand more about primitive ritual if we clarify further our ideas about secular rites. For us, individually, everyday symbolic enactment does several things. It provides a focussing mechanism, a method of mnemonics and a control for experience. To deal with focussing first, a ritual provides a frame. The marked off time or place alerts a special kind of expectancy, just as the oft-repeated 'Once upon a time' creates a mood receptive to fantastic tales. We can reflect on this framing function in small personal instances, for the least action is capable of carrying significance. Framing and boxing limit experience, shut in desired themes or shut out intruding ones. How many times is it necessary to fill a weekend case to find out how to exclude successfully all tokens of unwanted office life? One official file, packed in a weak moment, can spoil the whole effect of the holiday. I quote here Marion Milner on framing:

'. . . the frame marks off the different kind of reality that is within it from that which is outside it; but a temporal—spatial frame marks off the special kind of reality of a psycho-analytic session . . . makes possible the creative illusion called transference . . .' (1955)

She is discussing the technique of child analysis and mentions the locker in which the child patient keeps his play objects. It creates a kind of spatio-temporal frame which gives him continuity from one session to the next.

Not only does ritual aid us in selecting experiences for concentrated attention. It is also creative at the level of performance. For an external symbol can mysteriously help the co-ordination of brain and body. Actors' memoirs frequently recount cases in which a material symbol conveys effective power: the actor knows his part, he knows exactly how he wants to interpret it. But an intellectual knowing of what is to be done is not enough to produce the action. He tries continually and fails. One day some prop is passed to him, a hat or green umbrella, and with this symbol suddenly knowledge and intention are realised in the flawless performance.

The Dinka herdsman hurrying home to supper, knots a bundle of grass at the wayside, a symbol of delay. Thus he expresses outwardly his wish that the cooking may be delayed for his return. The rite holds no magic promise that he will now be in time for supper. He does not then dawdle home thinking that the action will itself be effective. He redoubles his haste. His action has not wasted time, for it has sharpened the focus of his attention on his wish to be in time (Lienhardt). The mnemonic action of rites is very familiar. When we tie knots in handkerchiefs we are not magicking our memory, but bringing it under the control of an external sign.

So ritual focusses attention by framing; it enlivens the memory and links the present with the relevant past. In all this it aids perception. Or rather, it changes perception because it changes the selective principles. So it is not enough to say that ritual helps us to experience more vividly what we would have experienced anyway. It is not merely like the visual aid which illustrates the verbal instructions for opening cans and cases. If it were just a kind of dramatic map or diagram of what is known it would always follow experience. But in fact ritual does not play this secondary role. It can come first in formulating experience. It can permit knowledge of what would otherwise not be known at all. It does not merely externalise experience, bringing it out into the light of day, but it modifies experience in so expressing it. This is true of language. There can be thoughts which have never been put into words. Once words have been framed the thought is changed and limited by the very words selected. So the speech has created something, a thought which might not have been the same.

There are some things we cannot experience without ritual. Events which come in regular sequences acquire a meaning from relation with others in the sequence. Without the full sequence individual elements become lost, imperceivable. For example, the days of the week, with their regular succession, names and distinctiveness: apart from their practical value in identifying the divisions of time, they each have meaning as part of a pattern. Each day has its own significance and if there are habits which establish the identity of a particular day, those regular observances have the effect of ritual. Sunday is not just a rest day. It is the day before Monday, and equally for Monday in relation to Tuesday. In a true sense we cannot experience Tuesday if for some reason we have not formally noticed that we have been through Monday. Going through one part of the pattern is a necessary procedure for being aware of the next part. Air travellers find that this applies to hours of the day and the sequence of meals. These are examples of symbols which are received and interpreted without having been intended. If we admit that they condition experience, so we must admit also that intended rituals in regular sequence can have this as one of their important functions.

Now we can turn to religious rites again. Durkheim was well aware that their effect is to create and control experience. It was his main preoccupation to study how religious ritual makes manifest to men their social selves and thus creates their society. But his thought was channelled into the English stream of anthropology by Radcliffe-Brown, who modified it. Thanks to Durkheim the primitive ritualist was no longer seen as a panto-mime magician. That was a notable advance on Frazer. Further-more, Radcliffe-Brown refused to separate religious ritual from secular ritual—another advance. Malinowski's magician became no different from any flag-waving patriot or superstitious salt-thrower, and these were treated alongside the Roman Catholic abstaining from meat or the Chinese putting rice on a grave. Ritual was no more mysterious or exotic.

In dropping both the words Sacred and Magic, Radcliffe-Brown seemed to restore the thread of continuity between secular and religious ritual. But unfortunately this failed to broaden the field of enquiry. For he wanted to use 'ritual' in a very narrow and special sense. It was to substitute for Durkheim's cult of the sacred and so be restricted to the enactment of soci-ally significant values (1939). Such-like constraints on the use of words are intended to help understanding. But so often they distort and confuse. Now we have got to the position in which Ritual replaces Religion in anthropologists' writings. It is used carefully and consistently to refer to symbolic action concerning the sacred. As a result the other, commoner kind of non-sacred ritual without religious efficacy has to be given another name if it is to be studied at all. So Radcliffe-Brown removed with one hand the barrier between sacred and secular, but put it back with the other. He also failed to follow up Durkheim's idea that ritual belongs within a social theory of knowledge, but treated it as part of a theory of action, taking on uncritically some assumptions about 'sentiments' current in the psychology of his day. Where there are common values, he said, rituals ex-press and focus attention on them. By rituals the necessary sentiments are generated to hold men to their roles. Childbirth taboos express to the Andaman Islanders the value of marriage and maternity and the danger to life in child labour. In war dances before a truce, the Andamans work off their sentiments of aggression. Food taboos instil sentiments of respect for seniority, and so on.

This approach is stultifying. Its main value is in requiring us to take taboos seriously because they express concern. But why the food taboos or visual or touch taboos should single out these particular foods or sights or contacts for avoidance is left unanswered. Radcliffe-Brown, somewhat in the spirit of Maimonides, implies that the question is silly, or that its answer is arbitrary. Even more unsatisfactory, we are left with little clue about people's concerns. It is obvious that death and childbirth should be a matter of concern. Thus Srinivas writing under the influence of Radcliffe-Brown says of Coorg avoidances and purifications:

'The pollution resulting from birth is milder than the pollution consequent on death. But in both cases pollution affects only the concerned kindred, and it is the means by which concern is defined and made known to everyone.' (1952, p. 102)

But he cannot apply the same reasoning to all pollutions. What sort of concern about bodily emissions, such as faeces or spittle, has to be defined and made known to everyone?

In the end the English received Durkheim's teaching when better field-work had raised understanding to the level of Durkheim's armchair insight. Lienhardt's whole discussion of Dinka religion is largely devoted to showing how rituals create and control experience. Writing of Dinka rain ceremonies, performed in the droughts of spring, he says:

'The Dinka themselves know, of course, when the rainy season is approaching . . . the point is of some importance for the correct appreciation of the spirit in which Dinka perform their regular ceremonies. In these their human symbolic action moves with the rhythm of the natural world around them, recreating that rhythm in moral terms and not merely attempting to coerce it into conformity with human desires.'

Lienhardt moves on in the same vein to sacrifices for health, for peace and to cancel the effects of incest. Finally he reaches the burial alive of Masters of the Fishing Spear, the rite by which the Dinka face and triumph over death itself. Throughout he insists on the rituals' function in modifying experience. Often it works retroactively. Officiants may solemnly deny the quarrels and misconduct which are the actual occasion of a sacrifice. This is not a cynical perjury at the altar itself. The object of the ritual is not to deceive God but to re-formulate past experience. By ritual and speech what has passed is restated so that what ought to have been prevails over what was, permanent good intention prevails over temporary aberration. When an act of incest has been committed, a sacrifice can alter the common descent of the pair and so expunge their guilt. The victim is cut in half alive, longitudinally through the sexual organs. So the common origin of the incestuous pair is symbolically negated. Similarly in peace-making ceremonies there are actions of blessing and purification as well as mimic battles:

> 'It seems that gesture without speech was enough to confirm in the external physical universe, an intention conceived interiorly in the moral. . . . The symbolic action in fact, mimes the total situation in which the parties in the feud know themselves to be including both their hostility and their disposition towards peace without which the ceremony could not be held. In this symbolic representation of their situation they control it, according to their will to peace, by transcending in symbolic action the only type of practical action (that is, continued hostilities) which for the Dinka follows from the situation of homicide.'

Later again (p. 291) he continues to hammer the point that ritual has as one of its objectives to control situations and to modify experience.

Only by establishing this point can he interpret the burial alive of the Dinka Spear Masters. Hence the fundamental principle is that certain men, closely in contact with Divinity, should not be seen to enter upon physical death.

> 'Their deaths are to be, or are to appear deliberate, and they are to be the occasion of a form of public celebration . . . the ceremonies in no way prevent the ultimate recognition of the ageing and physical death of those for whom they are performed. This death is recognised; but it is the public experience of it, for the survivors, which is deliberately modified by the performance of these ceremonies . . . the deliberately contrived death, though recognised as death, enables them to avoid admitting in this case the involuntary death which is the lot of ordinary men and beasts.'

The Master of the Fishing Spear does not kill himself. He requests a special form of death which is given by his people, for their own sake, not for his. If he were to die an ordinary death, the life of his people which is in his keeping, goes with him. His ritually contracted death separates his personal life from this public life. Everyone should rejoice, because on this occasion there is a social triumph over death.

Reading this account of Dinka attitudes to their rituals one gets the impression that the author is like a swimmer heading against a heavy tide. All the time he has to push aside the flow of arguments from simple-minded observers who have taken the ritual at its Aladdin-and-the-lamp face value. Of course Dinka hope that their rites will suspend the natural course of events. Of course they hope that rain rituals will cause rain, healing rituals avert death, harvest rituals produce crops. But instrumental efficacy is not the only kind of efficacy to be derived from their symbolic action. The other kind is achieved in the action itself, in the assertions it makes and the experience which bears its imprinting.

Once this has been forcefully spelled out for Dinka religious experience we cannot escape its truth. We can apply it even more fully to our own selves. First we should allow for the fact that very little of our ritual behaviour is enacted in the context of religion. Dinka culture is unified. Since all their major contexts of experience overlap and interpenetrate, nearly all their experience is religious, and so therefore is all their most important ritual. But our experiences take place in separate compartments and our rituals too. So we must treat the spring millinery and spring cleaning in our towns as renewal rites which focus and control experience as much as Swazi first fruit rituals.

When we honestly reflect on our busy scrubbings and cleanings in this light we know that we are not mainly trying to avoid disease. We are separating, placing boundaries, making visible statements about the home that we are intending to create out of the material house. If we keep the bathroom cleaning materials away from the kitchen cleaning materials and send the men to the downstairs lavatory and the women upstairs, we are essentially doing the same thing as the Bushman wife when she arrives at a new camp (Marshall Thomas, p. 41). She chooses where she will place her fire and then sticks a rod in the ground. This orientates the fire and gives it a right and left side. Thus the home is divided between male and female quarters.

We moderns operate in many different fields of symbolic

action. For the Bushman, Dinka and many primitive cultures, the field of symbolic action is one. The unity which they create by their separating and tidying is not just a little home, but a total universe in which all experience is ordered. Both we and the Bushmen justify our pollution avoidances by fear of danger. They believe that if a man sits on the female side his male virility will be weakened. We fear pathogenicity transmitted through micro-organisms. Often our justification of our own avoidances through hygiene is sheer fantasy. The difference between us is not that our behaviour is grounded on science and theirs on symbolism. Our behaviour also carries symbolic meaning. The real difference is that we do not bring forward from one context to the next the same set of ever more powerful symbols: our experience is fragmented. Our rituals create a lot of little sub-worlds, unrelated. Their rituals create one single, symbolically consistent universe. In the next two chapters we shall show what kinds of universes are produced when ritual and political needs work freely together.

Now to return to the question of efficacy. Mauss wrote of primitive society repaying itself with the false coin of magic. The metaphor of money admirably sums up what we want to assert of ritual. Money provides a fixed, external, recognisable sign for what would be confused, contradictable operations; ritual makes visible external signs of internal states. Money mediates transactions; ritual mediates experience, including social experience. Money provides a standard for measuring worth; ritual standardises situations, and so helps to evaluate them. Money makes a link between the present and the future, so does ritual. The more we reflect on the richness of the metaphor, the more it becomes clear that this is no metaphor. Money is only an extreme and specialised type of ritual.

In comparing magic with false currency Mauss was wrong. Money can only perform its role of intensifying economic interaction if the public has faith in it. If faith in it is shaken, the currency is useless. So too with ritual; its symbols can only have effect so long as they command confidence. In this sense all money, false or true, depends on a confidence trick. The test of money is whether it is acceptable or not. There is no false money except by contrast with another currency which has more total acceptability. So primitive ritual is like good money, not false money, so long as it commands assent.

Note that money can only generate economic activity by virtue of the feed-back from public confidence in it. What about ritual? What kind of effectiveness is generated by confidence in the power of its symbols? Using the analogy with coinage we can revive the question of magical efficacy. There are two possible views: either the power of magic is sheer illusion, or it is not. If it is not illusion, then symbols have power to work changes. Miracles apart, such a power could only work at two levels, that of individual psychology and that of social life. We know very well that symbols have power in social life; the analogy with currency provides an illustration. But has the Bank Rate anything to do with Shamanistic cures? Psychoanalysts claim to work cures by manipulating symbols. Has the confrontation with the subconscious anything to do with primitive spell-binding and loosing? I now cite two marvellous studies which must render scepticism out of date.

One is Turner's analysis of a Shamanistic cure, 'An Ndembu Doctor in Practice' (1964), which I summarise briefly. The technique of the cure was the famous one of cupping blood and seeming to extract a tooth from the body of the patient. The symptoms were palpitations, severe pains in the back and disabling weakness. The patient was also convinced that the other villagers were against him and withdrew completely from social life. Thus there was a mixture of physical and psychological disturbance. The doctor proceeded by finding out everything about the past history of the village, conducting seances in which everyone was encouraged to discuss their grudges against the patient, while he aired his grievances against them. Finally the blood-cupping treatment dramatically involved the whole village in a crisis of expectation that burst in the excitement of the extraction of the tooth from the bleeding, fainting patient. Joyfully they congratulated him on his recovery and themselves on their part in it. They had reason for joy since the long treatment had uncovered the main sources of tension in the village. In future the patient could play an acceptable role in their affairs. Dissident elements had been recognised and shortly left the village for good. The social structure was analysed and rearranged so that friction was, for the time, reduced.

In this absorbing study we are shown a case of skilful group therapy. The back-biting and envy of the villagers, symbolised by the tooth in the sick man's body, was dissolved in a wave of enthusiasm and solidarity. As he was cured of his physical symptoms they were all cured of social malaise. These symbols worked at the psycho-somatic level for the central figure, the sick man, and at the general psychological level for the villagers, in changing their attitudes, and at the sociological level in so far as the pattern of social statuses in the village was formally altered and in so far as some people moved in and others moved away as a result of the treatment.

In conclusion Turner says:

'Stripped of its supernatural guise, Ndembu therapy may well offer lessons for Western clinical practice. For relief might be given to many sufferers from neurotic illness if all those included in their social networks could meet together and publicly confess their ill will towards the patient and endure in their turn the recital of his grudges against them. But it is likely that nothing less than ritual sanctions for such behaviour and belief in the doctor's mystical powers could bring about such humility and compel people to display charity towards their suffering neighbour.'

This account of a Shamanistic cure points to the manipulation of the social situation as the source of its efficacy. The other enlightening study says nothing whatever about the social situation but concentrates on the direct power of the symbols to work upon the mind of the sufferer. Levi-Strauss (1949 & 1958), has analysed a Cuna Shaman's song which is chanted to relieve a difficult delivery in child birth. The doctor does not touch the patient. The incantation is to have its effect merely by recital. The song starts by describing the difficulties of the midwife and her appeal to the Shaman. Then the Shaman at the head of a band of protective spirits, sets out (in song) for the house of *Muu*, a power responsible for the foetus, which has captured the soul of the patient. The song describes the quest, the obstacles and dangers and victories of the Shaman's party until they finally give battle to *Muu* and her confederates. Once *Muu* is conquered and frees the captive soul, the labouring mother is delivered of her child and the song ends. The interest of the song is that the landmarks on the Shaman's journey to *Muu* are literally the vagina and womb of the pregnant woman, in the depths of which he finally fights for her victoriously. By repetition and minute detail, the song forces the patient to attend to an elaborate account of what has gone wrong in her labour. In one sense the body and internal organs of the patient are the theatre of action in the story, but by the transformation of the problem into a dangerous journey and battle with cosmic powers, by shuttling back and forth between the arena of the body and the arena of the universe, the Shaman is able to impose his view of the case. The patient's terror is focussed on the strength of mythic adversaries and her hopes of recovery fixed on the powers and ruses of the Shaman and his troups.

'The cure would consist then in making an emotional situation thinkable; and in making the mind accept pains which the body refuses to bear. It is of no importance that the mythology of the Shaman does not correspond to objective reality: the patient believes in it. The protective powers and the malevolent ones, the supernatural monsters and magic animals form part of a coherent system which underlies the native conception of the universe. The patient accepts them, or rather she has never doubted them. What she does not accept is this incoherent and arbitrary pain which is an intrusive element in her system. By appeal to the myth, the Shaman places it in a unified scheme where everything belongs. But the patient, having understood, does not resign herself: she gets better.'

Like Turner, Levi-Strauss also concludes his study with very pertinent suggestions for psychoanalysis.

These examples should be enough to shake a too complacent contempt of primitive religious beliefs. Not the absurd Ali Baba, but the magisterial figure of Freud is the model for appreciating the primitive ritualist. The ritual is creative indeed. More wonderful than the exotic caves and palaces of fairy tales, the magic of primitive ritual creates harmonious worlds with ranked and ordered populations playing their appointed parts. So far from being meaningless, it is primitive magic which gives meaning to existence. This applies as much to the negative as to the positive rites. The prohibitions trace the cosmic outlines and the ideal social order.

5

Primitive Worlds

'Now what are the characteristic marks of the sea-anemone', George Eliot muses, 'which entitle it to be removed from the hands of the botanist and placed in those of the zoologist?'

For us ambiguous species merely provoke essayists to elegant reflections. For Leviticus the rock badger or Syrian hyrax is unclean and abominable. Certainly it is an anomaly all right. It looks like an earless rabbit, has teeth like a rhino and the small hoofs on its toes seem to relate it to the elephant. But its existence does not threaten to bring the structure of our culture tumbling round our ears. Now that we have recognised and assimilated our common descent with apes nothing can happen in the field of animal taxonomy to rouse our concern. This is one reason why cosmic pollution is more difficult for us to understand than social pollutions of which we have some personal experience.

Another difficulty is our long tradition of playing down the difference between our own point of vantage and that of primitive cultures. The very real differences between 'us' and 'them' are made little of, and even the word 'primitive' is rarely used. Yet it is impossible to make any headway with a study of ritual pollution if we cannot face the question of why primitive culture is pollution-prone and ours is not. With us pollution is a matter of aesthetics, hygiene or etiquette, which only becomes grave in so far as it may create social embarrassment. The sanctions are social sanctions, contempt, ostracism, gossip, perhaps even police action. But in another large group of human societies the effects of pollution are much more wide ranging. A grave pollution is a religious offence. What is the basis of this difference? We cannot avoid the question and must attempt to phrase an objective, verifiable distinction between two types of culture, primitive and modern. Perhaps we Anglo-Saxons are more concerned to emphasise our sense of common humanity. We feel there is something discourteous in the term 'primitive' and so we avoid it and the whole subject too. Why else should Professor Herskovits have renamed the second edition of 'Primitive Economics' to 'Economic Anthropology' if his sophisticated West African friends had not expressed dislike of being lumped together with naked Fuegians and Aborigines under this general sign? Perhaps it is partly also in healthy reaction to early anthropology: 'Perhaps nothing so sharply differentiates the savage from the civilised man as the circumstance that the former observes tabus, the latter does not' (Rose, 1926, p. 111). No one can be blamed for wincing at a passage such as the following, though I do not know who would take it seriously:

> 'We know that the primitive man of today has mental equipment very different from that of the civilised man. It is much more fragmented, much more discontinuous, more "gestalt-free". Professor Jung once told me how, in his travels in the African bush, he had noticed the quivering eye-balls of his native guides: not the steady gaze of the European, but a darting restlessness of vision, due perhaps to the constant expectation of danger. Such eye movements must be co-ordinated with a mental alertness and a swiftly changing imagery that allows little opportunity for discursive reasoning, for contemplation and comparison.' (H. Read, 1955)

If this were written by a Professor of Psychology it might be significant, but it is not. I suspect that our professional delicacy in avoiding the term 'Primitive' is the product of secret convictions of superiority. The physical anthropologists have a similar problem. While they attempt to substitute 'ethnic group' for the word 'race', (see *Current Anthropology*, 1964) their terminological problems do not inhibit them from their task of distinguishing and classifying forms of human variation. But social anthropologists, to the extent that they avoid reflecting on the grand distinctions between human cultures, seriously impede their own work. So it is worth asking why the term 'primitive' should imply any denigration.

Part of our difficulty in England is that Levy-Bruhl, who first posed all the important questions about primitive cultures and their distinctiveness as a class, wrote in deliberate criticism of the English of his day, particularly of Frazer. Furthermore, Levy-Bruhl laid himself open to powerful counter-attack. Most text-books on comparative religion are emphatic about the mistakes he made, and say nothing about the value of the questions he asked. (For example, F. Bartlett, 1923, pp. 283-4, and P. Radin, 1956, pp. 230-1.) In my view he has not deserved such neglect.

Levy-Bruhl was concerned to document and to explain a peculiar mode of thought. He started (1922) with the problem set by an apparent paradox. On the one hand there were convincing reports of the high level of intelligence of Eskimo or Bushmen (or of other such hunters and gatherers, or primitive cultivators or herdsmen), and on the other hand reports of peculiar leaps made in their reasoning and interpretation of events which suggested that their thought followed very different paths from our own. He insisted that their alleged dislike of discursive reasoning is not due to intellectual incapacity but to highly selective standards of relevance which produce in them an 'insuperable indifference to matters bearing no apparent relation to those which interest them'. The problem then was to discover the principles of selection and of association which made the primitive culture favour explanation in terms of remote, invisible mystic agencies and to lack curiosity about the intermediate links in a chain of events. Sometimes Levy-Bruhl seems to be putting his problem in terms of individual psychology, but it is clear that he saw it as a problem of the comparison of cultures first and as a psychological one only in so far as individual psychology is affected by cultural environment. He was interested in analysing 'collective representations', that is standardised assumptions and categories, rather than in individual aptitudes. It is precisely on this score that he criticised Tylor and Frazer, who tried to explain primitive beliefs in terms of individual psychology, whereas he followed Durkheim in seeing collective representations as social phenomena, as common patterns of thought which are related to social institutions. In this he was undoubtedly right, but as his strength lay more in massive documentation than in analysis he was unable to apply his own precepts.

What Levy-Bruhl should have done, Evans-Pritchard has said, was to examine the variations in social structure and relate them to concomitant variations in the patterns of thought. Instead he contented himself with saying that all primitive people present uniform patterns of thought when contrasted with ourselves, and laid himself open to further criticism by seeming to make primitive cultures more mystical than they are and making civilised thought more rational than it is (Evans-Pritchard, *Levy-Bruhl's Theory of Primitive Mentality*). It seems that Evans-Pritchard himself was the first person to listen sympathetically to Levy-Bruhl and to direct his research to carrying Levy-Bruhl's problems into the more fruitful field which Levy-Bruhl himself missed. For his analysis of Azande witchcraft beliefs was exactly an exercise of this sort. It was the first study to describe a particular set of collective representations and to relate them intelligibly to social institutions (1937). Many studies have now ploughed lines parallel to this first furrow, so that from England and America a large body of sociological analysis of religions has vindicated Durkheim's insight. I say Durkheim's insight and not Levy-Bruhl's advisedly, for in so far as he contributed his own original slant to the matter, so Levy-Bruhl earned the just criticism of his reviewers. It was his idea to contrast primitive mentality with rational thought instead of sticking to the problem adumbrated by the master. If he had stayed with Durkheim's view of the problem he would not have been led into the confusing contrast of mystical with scientific thought, but would have compared primitive social organisation with complex modern social organisation and perhaps have done something useful towards elucidating the difference between organic and mechanical solidarity, between two types of social organisation which Durkheim saw to underlie differences in beliefs.

Since Levy-Bruhl the general tendency in England has been to treat each culture studied as wholly *sui generis,* a unique and more or less successful adaptation to a particular environment (see Beattie, 1960, p. 83, 1964, p. 272). Evans-Pritchard's criticism that Levy-Bruhl treated primitive cultures as if they were more uniform than they really are has stuck. But it is vital now to take up this matter again. We cannot understand sacred contagion unless we can distinguish a class of cultures in which pollution ideas flourish from another class of cultures, including our own, in which they do not. Old Testament scholars do not hesitate to enliven their interpretations of Israelite culture by comparison with primitive cultures. Psychoanalysts since Freud, and metaphysicians since Cassirer are not backward in drawing general comparisons between our present civilisation and others very different. Nor can anthropologists do without such general distinctions.

The right basis for comparison is to insist on the unity of human experience and at the same time to insist on its variety, on the differences which make comparison worth while. The only way to do this is to recognise the nature of historical progress and the nature of primitive and of modern society. Progress means differentiation. Thus primitive means undifferentiated; modern means differentiated. Advance in technology involves differentiation in every sphere, in techniques and materials, in productive and political roles.

We could, theoretically, construct a rough gradient along which different economic systems would lie according to the degree to which they have developed specialised economic institutions. In the most undifferentiated economies roles in the productive system are not allocated by market considerations and there are few specialised labourers or craftsmen. A man does what work he does as part of performing his role as, say, son or brother or head of family. The same goes for the processes of distribution. As there is no labour exchange, so there is no supermarket. Individuals get their share of the community's product in virtue of their membership; their age, sex, seniority, their relationship to others. The patterns of status are etched by grooves of obligatory gift-making, along which rights to wealth are channelled.

Unfortunately for economic comparison there are many societies, small in scale, based on primitive techniques, which are not organised in this way, but rather on principles of market competition (see Pospisil). However, development in the political sphere lends itself very satisfactorily to the pattern I wish to introduce. There are not, in the most small-scale type of society, any specialised political institutions. Historical progress is marked by the development of diverse judicial, military, police, parliamentary and bureaucratic institutions. So it is easy enough to trace what internal differentiation would mean for social institutions.

On the face of it the same process should be traceable in the intellectual sphere. It seems unlikely that institutions should diversify and proliferate without a comparable movement in the realm of ideas. Indeed we know that it does not happen. Great steps separate the historical development of the Hadza in Tanganyikan forests, who still never have occasion to count beyond four, from that of West Africans who for centuries have reckoned fines and taxes in thousands of cowries. Those of us who have not mastered modern techniques of communication such as the language of mathematics or of computers can put ourselves in the Hadza class compared with the ones who have become articulate in these media. We know only too well the educational burden our own civilisation carries in the form of specialised compartments of learning. Obviously the demand for special expertise and the education for providing it create cultural environments in which certain kinds of thinking can flourish and others cannot. Differentiation in thought patterns goes along with differentiated social conditions.

From this basis it ought to be straightforward to say that in the realm of ideas there are differentiated thought systems which contrast with undifferentiated ones, and leave it at that. But the trap is just here. What could be more complex, diversified and elaborate than the Dogon cosmology? Or the Australian Murinbata cosmology, or the cosmology of Samoa, or of Western Pueblo Hopi for that matter? The criterion we are looking for is not in elaborateness and sheer complication of ideas.

There is only one kind of differentiation in thought that is relevant, and that provides a criterion that we can apply equally to different cultures and to the history of our own scientific ideas. That criterion is based on the Kantian principle that thought can only advance by freeing itself from the shackles of its own subjective conditions. The first Copernican revolution, the discovery that only man's subjective viewpoint made the sun seem to revolve round the earth, is continually renewed. In our own culture mathematics first and later logic, now history, now language and now thought processes themselves and even knowledge of the self and of society, are fields of knowledge progressively freed from the subjective limitations of the mind. To the extent to which sociology, anthropology and psychology are possible in it, our own type of culture needs to be distinguished from others which lack this self-awareness and conscious reaching for objectivity.

Radin interprets the Trickster myth of the Winnebago Indians on lines which serve to illustrate this point. Here is a primitive parallel to Teilhard de Chardin's theme that the movement of evolution has been towards ever-increasing complexification and self-awareness.

These Indians lived technically, economically and politically in the most simple undifferentiated conditions. Their myth contains their profound reflections on the whole subject of differentiation. The trickster starts as an unselfconscious, amorphous being. As the story unfolds he gradually discovers his own identity, gradually recognises and controls his own anatomical parts; he oscillates between female and male, but eventually fixes his own male sexual role; and finally learns to assess his environment for what it is. Radin says in his preface:

> 'He wills nothing consciously. At all times he is constrained to behave as he does from impulses over which he has no control . . . he is at the mercy of his passions and appetites . . . possesses no defined and well-fixed form . . . primarily an inchoate being of indeterminate proportions, a figure foreshadowing the shape of man. In this version he possesses intestines wrapped around his body and an equally long penis, likewise wrapped round his body with his scrotum on top of it.'

Two examples of his strange adventures will illustrate this theme. Trickster kills a buffalo and is butchering it with a knife in his right hand:

> 'In the midst of all these operations suddenly his left arm grabbed the buffalo. "Give that back to me, it is mine! Stop that or I will use my knife on you!" So spoke the right arm. "I will cut you to pieces, that is what I will do to you," continued the right arm. Thereupon the left arm released its hold. But shortly after, the left arm again grabbed hold of the right arm . . . again and again this was repeated. In this manner did Trickster make both his arms quarrel. That quarrel soon turned into a vicious fight and the left arm was badly cut up. . . .'

In another story Trickster treats his own anus as if it could act as an independent agent and ally. He had killed some ducks and before going to sleep he tells his anus to keep guard over the meat. While he is asleep some foxes draw near:

> 'When they came close, much to their surprise however, gas was expelled from somewhere. "Pooh" was the sound made. "Be careful! He must be awake", so they ran back. After a while one of them said "Well, I guess he is asleep now. That was

only a bluff. He is always up to some tricks. So again they approached the fire. Again gas was expelled and again they ran back. Three times this happened . . . Then louder, still louder, was the sound of gas expelled. "Pooh! Pooh! Pooh!" Yet they did not run away. On the contrary they now began to eat the roasted pieces of duck. . . .'

When Trickster woke up and saw the duck gone:

'. . . "Oh, you too, you despicable object, what about your be-haviour? Did I not tell you to watch this fire? You shall remember this! As a punishment for your remissness, I will burn your mouth so that you will not be able to use it!" So he took a piece of burning wood and burned the mouth of his anus . . . and cried out of pain he was inflicting on himself.'

Trickster begins, isolated, amoral and unselfconscious, clumsy, ineffectual, an animal-like buffoon. Various episodes prune down and place more correctly his bodily organs so that he ends by looking like a man. At the same time he begins to have a more consistent set of social relations and to learn hard lessons about his physical environment. In one important episode he mistakes a tree for a man and responds to it as he would to a person until eventually he discovers it is a mere inanimate thing. So gradually he learns the functions and limits of his being.

I take this myth as a fine poetic statement of the process that leads from the early stages of culture to contemporary civilisa-tion, differentiated in so many ways. The first type of culture is not pre-logical, as Levy-Bruhl unfortunately dubbed it, but pre-Copernican. Its world revolves round the observer who is trying to interpret his experiences. Gradually he separates him-self from his environment and perceives his real limitations and powers. Above all this pre-Copernican world is personal. Trick-ster speaks to creatures, things and parts of things without dis-crimination as if they were animate, intelligent beings. This personal universe is the kind of universe that Levy-Bruhl describes. It is also the primitive culture of Tylor and the animist culture of Marett, and the mythological thought of Cassirer.

In the next few pages I am going to press as hard as I can the analogy between primitive cultures and the early episodes of the Trickster myth. I will try to present the characteristic areas of non-differentiation which define the primitive world view. I shall develop the impression that the primitive world view is subjective and personal, that different modes of existence are confused, that the limitations of man's being are not known. This is the view of primitive culture which was accepted by Tylor and Frazer and which posed the problems of primitive mentality. I shall then try to show how this approach distorts the truth.

First, this world view is man-centred in the sense that explanations of events are couched in notions of good and bad fortune, which are implicitly subjective notions ego-centred in reference. In such a universe the elemental forces are seen as linked so closely to individual human beings that we can hardly speak of an external, physical environment. Each individual carries within himself such close links with the universe that he is like the centre of a magnetic field of force. Events can be explained in terms of his being what he is and doing what he has done. In this world it makes good sense for Thurber's fairy tale king to complain that falling meteors are being hurled at himself, and for Jonah to come forward and confess that he is the cause of a storm. The distinctive point here is not whether the working of the universe is thought to be governed by spiritual beings or by impersonal powers. That is hardly relevant. Even powers which are taken to be thoroughly impersonal are held to be reacting directly to the behaviour of individual humans.

A good example of belief in anthropocentric powers is the !Kung Bushmen belief in *N!ow*, a force thought to be responsible for meteorological conditions at least in the Nyae-Nyae area of Bechuanaland. *N!ow* is an impersonal, amoral force, definitely a thing and not a person. It is released when a hunter who has one kind of physical make-up kills an animal which has the corresponding element in its own make-up. The actual weather at any time is theoretically accounted for by the complex interaction of different hunters with different animals (Marshall). This hypothesis is attractive and one feels it must be intellectually satisfying since it is a view which is theoretically capable of being verified and yet no serious testing would ever be practical.

To illustrate further the man-centred universe I quote from what Father Tempels says of Luba philosophy. He has been criticised for implying that what he says so authoritatively from his intimate knowledge of Luba thought applies to all the Bantu.

But I suspect that in its broad lines his view on Bantu ideas of vital force applies not merely to all the Bantu, but much more widely. It probably applies to the whole range of thought which I am seeking to contrast with modern differentiated thought in European and American cultures.

For the Luba, he says, the created universe is centred on man (pp. 43-5). The three laws of vital causality are:

(1) that a human (living or dead) can directly reinforce or diminish the being (or force) of another human
(2) that the vital force of a human can directly influence in-ferior force-beings (animal, vegetable or mineral)
(3) that a rational being (spirit, dead or living human) can act indirectly on another by communicating his vital influence to an intermediary inferior force.

Of course there are very many different forms which the idea of a man-centred universe may take. Inevitably ideas of how men affect other men must reflect political realities. So ultimately we shall find that these beliefs in man-centred control of the en-vironment vary according to the prevailing tendencies in the political system (see Chapter 6). But in general we can dis-tinguish beliefs which hold that all men are equally involved with the universe from beliefs in the special cosmic powers of selected individuals. There are beliefs about destiny which are thought to apply universally to all men. In the culture of Homeric literature it was not certain outstanding individuals whose destiny was the concern of the gods, but all and each whose personal fate was spun on the knees of the gods and woven for good or ill with the fates of others. Just to take one contemporary example, Hinduism today teaches, as it has for centuries, that for each individual the precise conjunction of the planets at the time he was born signifies much for his personal good or ill-fortune. Horoscopes are for everybody. In both these instances, though the individual can be warned by diviners about what is in store for him, he cannot change it radically, only soften a little the hard blows, defer or abandon hopeless desires, be alert to the opportunities that will lie in his path.

Other ideas about the way in which the individual's fortune is bound up with the cosmos may be more pliable. In many parts of West Africa today, the individual is held to have a complex personality whose component parts act like separate persons. One part of the personality speaks the life-course of the individual before he is born. After birth, if the individual strives for success in a sphere which has been spoken against, his efforts will always be in vain. A diviner can diagnose this spoken destiny as cause of his failures and can then exorcise his prenatal choice. The nature of his pre-destined failure which a man has to take account of varies from one West African society to another. Among the Tallensi in the Ghana hinterland the conscious personality is thought to be amiable and uncompetitive. His unconscious element which spoke his destiny before birth is liable to be diagnosed as over-aggressive and rivalrous, and so makes him a misfit in a system of controlled statuses. By contrast the Ijo of the Niger Delta, whose social organisation is fluid and competitive, take the conscious component of the self to be full of aggression, desire to compete and to excel. In this case it is the unconscious self which may be pre-destined to failure because it chose obscurity and peace. Divination can discover the discrepancy of aims within the person, and ritual can put it right. (Fortes, 1959; Horton, 1961.)

These examples point to another lack of differentiation in the personal world view. We saw above that the physical environment is not clearly thought of in separate terms, but only with reference to the fortunes of human selves. Now we see that the self is not clearly separated as an agent. The extent and limits of its autonomy are not defined. So the universe is part of the self in a complementary sense, seen from the angle of the individual's idea, not this time of nature, but of himself.

The Tallensi and Ijo ideas about the multiple warring personalities in the self seem to be more differentiated than the Homeric Greek idea. In these West African cultures the binding words of destiny are spoken by part of the individual himself. Once he knows what he has done he can repudiate his earlier choice. In Ancient Greece the self was seen as a passive victim of external agents:

'In Homer one is struck by the fact that his heroes with all their magnificent vitality and activity feel themselves at every turn not free agents but passive instruments or victims of other powers . . . a man felt that he could not help his own emotions. An idea, an emotion, an impulse came to him; he acted and presently rejoiced or lamented. Some god had inspired him or blinded him. He prospered, then was poor, perhaps enslaved;

he wasted away with disease, or died in battle. It was divinely ordained, his portion apportioned long before. The prophet or diviner might discover it in advance; the plain man knew a little about omens and merely seeing his shaft hit its mark or the enemy prevailing, concluded that Zeus had assigned defeat to himself and his comrades. He did not wait to fight further but fled.' (Onians, 1951, p. 302)

The pastoral Dinka living in the Sudan similarly are said not to distinguish the self as an independent source of action and of reaction. They do not reflect on the fact that they themselves react with feelings of guilt and anxiety and that these feelings initiate other states of mind. The self acted upon by emotions they portray by external powers, spiritual beings who cause misfortune of various kinds. So in an effort to do justice to the complex reality of the self's interaction within itself the Dinka universe is peopled with dangerous personal extensions to the self. This is almost exactly how Jung described the primitive world view when he said:

'An unlimited amount of what we now consider an integral part of our own psychic being disports itself merrily for the primitive in projections reaching far and wide.' (p. 74)

I give one more example of a world in which all individuals are seen as personally linked with the cosmos to show how varied these linkages can be. Chinese culture is dominated by the idea of harmony in the universe. If an individual can place himself to ensure the most harmonious relationship possible, he can hope for good fortune. Misfortune may be attributed to lack of just such a happy alignment. The influence of the waters and the airs, called Feng Shwe, will bring him good fortune if his house and his ancestors' graves are well placed. Professional geomancers can divine the causes of his misfortune and he can then rearrange his home or his parental graves to better effect. Dr. Freedman in his forthcoming book holds that geomancy has an important place in Chinese beliefs alongside ancestor worship. The fortune which a man can manipulate thus by geomantic skills has no moral implications; but ultimately it must be brought to terms with the reward of merit which in the same set of beliefs is meted out by heaven. Finally then, the whole universe is interpreted as tied in its detailed workings to the lives of human persons. Some individuals are more successful in dealing with Feng Shwe than others, just as some Greeks have a more splendid fate decreed for them and some West Africans a spoken destiny more committed to success.

Sometimes it is only marked individuals and not all humans who are significant. Such marked individuals draw lesser men in their wake, whether their endowment is for good or evil fortune. For the ordinary man in the street, not endowed himself, the practical problem is to study his fellow men and discover whom among them he ought to avoid or follow.

In all the cosmologies we have mentioned so far, the lot of individual humans is thought to be affected by power inhering in themselves or in other humans. The cosmos is turned in, as it were, on man. Its transforming energy is threaded on to the lives of individuals so that nothing happens in the way of storms, sickness, blights or droughts except in virtue of these personal links. So the universe is man-centred in the sense that it must be interpreted by reference to humans.

But there is a quite other sense in which the primitive undifferentiated world view may be described as personal. Persons are essentially not things. They have wills and intelligence. With their wills they love, hate and respond emotionally. With their intelligence they interpret signs. But in the kind of universe I am contrasting with our own world view, things are not clearly distinct from persons. Certain kinds of behaviour characterise person to person relations. First, persons communicate with one another by symbols in speech, gesture, rite, gift and so on. Second, they react to moral situations. However impersonally the cosmic forces may be defined, if they seem to respond to a person-to-person style of address their quality of thing is not fully differentiated from their personality. They may not be persons but nor are they entirely things.

Here there is a trap to avoid. Some ways of talking about things might seem to the naïve observer to imply personality. Nothing can necessarily be inferred about beliefs from purely linguistic distinctions or confusions. For instance a Martian anthropologist might come to the wrong conclusion on overhearing an English plumber asking his mate for the male and female parts of plugs. To avoid falling into linguistic pitfalls, I confine my interests to the kind of behaviour which is supposed to produce a response from allegedly impersonal forces.

It may not be at all relevant here that the Nyae-Nyae Bushmen attribute male and female character to clouds, any more than it is relevant that we use 'she' for cars and boats. But it may be relevant that the pygmies of the Ituri forest, when misfortune befalls, say that the forest is in a bad mood and go to the trouble of singing to it all night to cheer it up, and that they then expect their affairs to prosper (Turnbull). No European mechanic in his senses would hope to cure engine trouble by serenade or curse.

So here is another way in which the primitive, undifferentiated universe is personal. It is expected to behave as if it was intelligent, responsive to signs, symbols, gestures, gifts, and as if it could discern between social relationships.

The most obvious example of impersonal powers being thought responsive to symbolic communication is the belief in sorcery. The sorcerer is the magician who tries to transform the path of events by symbolic enactment. He may use gestures or plain words in spells or incantations. Now words are the proper mode of communication between persons. If there is an idea that words correctly said are essential to the efficacy of an action, then, although the thing spoken to cannot answer back, there is a belief in a limited kind of one-way verbal communication. And this belief obscures the clear thing-status of the thing being addressed. A good example is the poison used for the oracular detection of witches in Zandeland (Evans-Pritchard, 1937). The Azande themselves brew their poison from bark. It is not said to be a person but a thing. They do not suppose there is a little man inside which works the oracle. Yet for the oracle to work the poison must be addressed aloud, the address must convey the question unequivocally and, to eliminate error of interpretation, the same question must be put in reverse form in the second round of consultation. In this case not only does the poison hear and understand the words, but it has limited powers of reply. Either it kills the chicken or it does not. It can only give yes and no answers. It cannot initiate a conversation or conduct an unstructured interview. Yet this limited response to questioning radically modifies its thing-status in the Azande universe. It is not an ordinary poison, but more like a captive interviewee filling in a survey questionnaire with crosses and ticks.

The Golden Bough is full of examples of belief in an impersonal universe which, nevertheless, listens to speech and responds to it one way or another. So are modern field-workers' reports. Stanner says: 'Most of the choir and furniture of heaven and earth are regarded by the Aborigines as a vast sign system. Anyone who understandingly has moved in the Australian bush with aborigine associates becomes aware of the fact. He moves, not in a landscape but in a humanised realm saturated with significations.'

Finally there are the beliefs which imply that the impersonal universe has discernment. It may discern between fine nuances in social relations, such as whether the partners in sexual intercourse are related within prohibited degrees, or between less fine ones such as whether a murder has been committed on a fellow-tribesman or on a stranger, or whether a woman is married or not. Or it may discern secret emotions hidden in men's breasts. There are many examples of implied discernment of social status. The hunting Cheyenne thought that the buffaloes who provide their main livelihood were affected by the rotten smell of a man who had murdered a fellow-tribesman and they moved away, thus endangering the survival of the tribe. The buffalo were not supposed to react to the smell of murder of a foreigner. The Australian Aborigines of Arnhemland conclude their fertility and initiation ceremonies with ceremonial copulation, believing that the rite is more efficacious if sexual intercourse takes place between persons who are at other times strictly prohibited (Berndt, p. 49). The Lele believe that a diviner who has had sexual intercourse with the wife of his patient, or whose patient has had sexual intercourse with his wife, cannot heal him, because the medicine intended to heal would kill. This result is not dependent on any intention or knowledge on the part of the doctor. The medicine itself is thought to react in this discriminating way. Furthermore, the Lele believe that if a cure is effected and the patient omits to pay his healer promptly for his services, early relapse or even a more fatal complication of the illness will result. So Lele medicine, by implication, is credited with discerning debt as well as secret adultery. Even more intelligent is the vengeance magic bought by the Azande which detects unerringly the witch responsible for a given death, and does capital justice on him. So impersonal elements in the universe are credited with discrimination which enables them to intervene in human affairs and uphold the moral code.

In this sense the universe is apparently able to make judgments on the moral value of human relations and to act accordingly. *Malweza*, among the Plateau Tonga in Northern Rhodesia, is a misfortune which afflicts those who commit certain specific offences against the moral code. Those offences are in general of a kind against which ordinary punitive sanctions cannot be applied. For example, homicide within the group of matrilineal kinsmen cannot be avenged because the group is organised to avenge the murder of its members by outsiders (Colson, p. 107). *Malweza* punishes offences which are inaccessible to ordinary sanctions.

To sum up, a primitive world view looks out on a universe which is personal in several different senses. Physical forces are thought of as interwoven with the lives of persons. Things are not completely distinguished from persons and persons are not completely distinguished from their external environment. The universe responds to speech and mime. It discerns the social order and intervenes to uphold it.

I have done my best to draw from accounts of primitive cultures a list of beliefs which imply lack of differentiation. The materials I have used are based on modern fieldwork. Yet the general picture closely accords with that accepted by Tylor or Marett in their discussions of primitive animism. They are the kind of beliefs from which Frazer inferred that the primitive mind confused its subjective and objective experiences. They are the same beliefs which provoked Levy-Bruhl to reflect on the way that collective representations impose a selective principle on interpretation. The whole discussion of these beliefs has been haunted by obscure psychological implications.

If these beliefs are presented as the result of so many failures to discriminate correctly they evoke to a startling degree the fumbling efforts of children to master their environment. Whether we follow Klein or Piaget, the theme is the same; confusion of internal and external, of thing and person, self and environment, sign and instrument, speech and action. Such confusions may be necessary and universal stages in the passage of the individual from the chaotic, undifferentiated experience of infancy to intellectual and moral maturity.

So it is important to point out again, as has often been said before, that these connections between persons and events which characterise the primitive culture do not derive from failure to differentiate. They do not even necessarily express the thoughts

88

of individuals. It is quite possible that individual members of such cultures hold very divergent views on cosmology. Vansina recalls affectionately three very independent thinkers he encountered among the Bushong, who liked to expound their personal philosophies to him. One old man had come to the conclusion that there was no reality, that all experience is a shifting illusion. The second had developed a numerological type of metaphysics, and the last had evolved a cosmological scheme of great complexity which no one understood but himself (1964). It is misleading to think of ideas such as destiny, witchcraft, *mana*, magic as part of philosophies, or as systematically thought out at all. They are not just linked to institutions, as Evans-Pritchard put it, but they are institutions—every bit as much as Habeas Corpus or Hallow-e'en. They are all compounded part of belief and part of practice. They would not have been recorded in the ethnography if there were no practices attached to them. Like other institutions they are both resistant to change and sensitive to strong pressure. Individuals can change them by neglect or by taking an interest.

If we remember that it is a practical interest in living and not an academic interest in metaphysics which has produced these beliefs, their whole significance alters. To ask an Azande whether the poison oracle is a person or a thing is to ask a kind of nonsensical question which he would never pause to ask himself. The fact that he addresses the poison oracle in words does not imply any confusion whatever in his mind between things and persons. It merely means that he is not striving for intellectual consistency and that in this field symbolic action seems appropriate. He can express the situation as he sees it by speech and mime, and these ritual elements have become incorporated into a technique which, to many intents and purposes is like programming a problem through a computer. I think that this is something argued by Radin in 1927 and by Gellner (1962) when he points to the social function of incoherences in doctrines and concepts.

Robertson Smith first tried to draw attention away from beliefs considered as such, to the practices associated with them. And much other testimony has piled up since on the strictly practical limitation on the curiosity of individuals. This is not a peculiarity of primitive culture. It is true of 'us' as much as of 'them', in so far as 'we' are not professional philosophers. As

business man, farmer, housewife no one of us has time or in-
clination to work out a systematic metaphysics. Our view of the
world is arrived at piecemeal, in response to particular practical
problems.

In discussing Azande ideas about witchcraft Evans-Pritchard
insists on this concentration of curiosity on the singularity of
an individual event. If an old and rotten granary falls down and
kills someone sitting in its shadow, the event is ascribed to
witchcraft. Azande freely admit that it is in the nature of old
and rotten granaries to collapse, and they admit that if a person sits for several hours under its shadow, day after day, he
may be crushed when it falls. The general rule is obvious and
not an interesting field for speculation. The question that
interests them is the emergence of a unique event out of the
meeting point of two separate sequences. There were many hours
when no one was sitting under that granary and when it might
have collapsed harmlessly, killing no one. There were many
hours when other people were seated by it, who might have been
victims when it fell, but who happened not to be there. The
fascinating problem is why it should have fallen just when it
did, just when so-and-so and no one else was sitting there. The
general regularities of nature are observed accurately and finely
enough for the technical requirements of Azande culture. But
when technical information has been exhausted, curiosity turns
instead to focus on the involvement of a particular person with
the universe. Why did it have to happen to him? What can he
do to prevent misfortune? Is it anyone's fault? This applies, of
course, to a theistic world view. As with witchcraft only certain
questions are answered by reference to spirits. The regular pro-
cession of the seasons, the relation of cloud to rain and rain to
harvest, of drought to epidemic and so on, is recognised. They
are taken for granted as the back-drop against which more personal and pressing problems can be solved. The vital questions
in any theistic world-view are the same as for the Azande: why
did this farmer's crops fail and not his neighbour's? Why did
this man get gored by a wild buffalo and not another of his
hunting party? Why did this man's children or cows die? Why
me? Why today? What can be done about it? These insistent
demands for explanation are focussed on an individual's concern
for himself and his community. We now know what Durkheim
knew, and what Frazer, Tylor and Marett did not. These ques-

tions are not phrased primarily to satisfy man's curiosity about the seasons and the rest of the natural environment. They are phrased to satisfy a dominant social concern, the problem of how to organise together in society. They can only be answered, it is true, in terms of man's place in nature. But the metaphysic is a by-product, as it were, of the urgent practical concern. The anthropologist who draws out the whole scheme of the cosmos which is implied in these practices does the primitive culture great violence if he seems to present the cosmology as a systematic philosophy subscribed to consciously by individuals. We can study our own cosmology—in a specialised department of astronomy. But primitive cosmologies cannot rightly be pinned out for display like exotic lepidoptera, without distortion to the nature of a primitive culture. In a primitive culture the technical problems have been more or less settled for generations past. The live issue is how to organise other people and oneself in relation to them; how to control turbulent youth, how to soothe disgruntled neighbours, how to gain one's rights, how to prevent usurpation of authority, or how to justify it. To serve these practical social ends all kinds of beliefs in the omniscience and omnipotence of the environment are called into play. If social life in a particular community has settled down into any sort of constant form, social problems tend to crop up in the same areas of tension or strife. And so as part of the machinery for resolving them, these beliefs about automatic punishment, destiny, ghostly vengeance and witchcraft crystallise in the institutions. So the primitive world view which I have defined above is rarely itself an object of contemplation and speculation in the primitive culture. It has evolved as the appanage of other social institutions. To this extent it is produced indirectly, and to this extent the primitive culture must be taken to be unaware of itself, unconscious of its own conditions.

In the course of social evolution institutions proliferate and specialise. The movement is a double one in which increased social control makes possible greater technical developments and the latter opens the way to increased social control again. Finally we find ourselves in the modern world where economic interdependence is carried to the highest pitch reached by mankind so far. One inevitable by-product of social differentiation is social awareness, self-consciousness about the processes of communal life. And with differentiation go special forms of social coercion, special monetary incentives to conform, special types of punitive sanctions, specialised police and overseers and progress men scanning our performance, and so on, a whole paraphernalia of social control which would never be conceivable in small-scale undifferentiated economic conditions. This is the experience of organic solidarity which makes it so hard for us to interpret the efforts of men in primitive society to overcome the weakness of their social organisation. Without forms filled in triplicate, without licences and passports and radio-police cars they must somehow create a society and commit men and women to its norms. I hope I have now shown why Levy-Bruhl was mistaken in comparing one type of thought with another instead of comparing social institutions.

We can also see why Christian believers, Moslems and Jews are not to be classed as primitive on account of their beliefs. Nor necessarily Hindus, Buddhists or Mormons, for that matter. It is true that their beliefs are developed to answer the questions 'Why did it happen to me: Why now?' and the rest. It is true that their universe is man-centred and personal. Perhaps in entertaining metaphysical questions at all these religions may be counted anomalous institutions in the modern world. For unbelievers may leave such problems aside. But this in itself does not make of believers promontories of primitive culture sticking out strangely in a modern world. For their beliefs have been phrased and rephrased with each century and their inter-meshing with social life cut loose. The European history of ecclesiastical withdrawal from secular politics and from secular intellectual problems to specialised religious spheres is the history of this whole movement from primitive to modern.

Finally we should revive the question of whether the word 'primitive' should be abandoned. I hope not. It has a defined and respected sense in art. It can be given a valid meaning for technology and possibly for economics. What is the objection to saying that a personal, anthropocentric, undifferentiated world-view characterises a primitive culture? The only source of objection could be from the notion that it has a pejorative sense in relation to religious beliefs which it does not carry in technology and art. There may be something in this for a certain section of the English-speaking world.

The idea of a primitive economy is slightly romantic. It is true that we are materially and technically incomparably better equipped, but no one would frankly base a cultural distinction on purely materialist grounds. The facts of relative poverty and wealth are not in question. But the idea of the primitive economy is one which handles goods and services without the intervention of money. So the primitives have the advantage over us in that they encounter economic reality direct, while we are always being deflected from our course by the complicated, unpredictable and independent behaviour of money. But on this basis, when it comes to the spiritual economy, we seem to have the advantage. For their relation to their external environment is mediated by demons and ghosts whose behaviour is complicated and unpredictable, while we encounter our environment more simply and directly. This latter advantage we owe to our wealth and material progress which has enabled other developments to take place. So, on this reckoning, the primitive is ultimately at a disadvantage both in the economic and spiritual field. Those who feel this double superiority are naturally inhibited from flaunting it and this is presumably why they prefer not to distinguish primitive culture at all.

Continentals seem to have no such squeamishness. *'Le primitif'* enjoys honour in the pages of Leenhard, Levi-Strauss, Ricoeur and Eliade. The only conclusion that I can draw is that they are not secretly convinced of superiority, and are intensely appreciative of forms of culture other than their own.

6

Powers and Dangers

GRANTED THAT DISORDER SPOILS PATTERN; it also provides the materials of pattern. Order implies restriction; from all possible materials, a limited selection has been made and from all possible relations a limited set has been used. So disorder by implication is unlimited, no pattern has been realised in it, but its potential for patterning is indefinite. This is why, though we seek to create order, we do not simply condemn disorder. We recognise that it is destructive to existing patterns; also that it has potentiality. It symbolises both danger and power.

Ritual recognises the potency of disorder. In the disorder of the mind, in dreams, faints and frenzies, ritual expects to find powers and truths which cannot be reached by conscious effort. Energy to command and special powers of healing come to those who can abandon rational control for a time. Sometimes an Andaman Islander leaves his band and wanders in the forest like a madman. When he returns to his senses and to human society he has gained occult power of healing (Radcliffe Brown, 1933, p. 139). This is a very common notion, widely attested. Webster in his chapter on the Making of a Magician (*The Sociological Study of Magic*), gives many examples. I also quote the Ehanzu, a tribe in the central region of Tanzania, where one of the recognised ways of acquiring a diviner's skill is by going mad in the bush. Virginia Adam, who worked among this tribe, tells me that their ritual cycle culminates in annual rain rituals. If at the expected time rain fails, people suspect sorcery. To undo the effects of sorcery they take a simpleton and send him wandering into the bush. In the course of his wanderings he unknowingly destroys the sorcerer's work.

94

In these beliefs there is a double play on inarticulateness. First there is a venture into the disordered regions of the mind. Second there is the venture beyond the confines of society. The man who comes back from these inaccessible regions brings with him a power not available to those who have stayed in the control of themselves and of society.

This ritual play on articulate and inarticulate forms is crucial to understanding pollution. In ritual form it is treated as if it were quick with power to maintain itself in being, yet always liable to attack. Formlessness is also credited with powers, some dangerous, some good. We have seen how the abominations of Leviticus are the obscure unclassifiable elements which do not fit the pattern of the cosmos. They are incompatible with holiness and blessing. The play on form and formlessness is even more clear in the rituals of society.

First, consider beliefs about persons in a marginal state. These are people who are somehow left out in the patterning of society, who are placeless. They may be doing nothing morally wrong, but their status is indefinable. Take, for example, the unborn child. Its present position is ambiguous, its future equally. For no one can say what sex it will have or whether it will survive the hazards of infancy. It is often treated as both vulnerable and dangerous. The Lele regard the unborn child and its mother as in constant danger, they also credit the unborn child with capricious ill-will which makes it a danger to others. When pregnant, a Lele woman tries to be considerate about not approaching sick persons lest the proximity of the child in her womb causes coughing or fever to increase.

Among the Nyakyusa a similar belief is recorded. A pregnant woman is thought to reduce the quantity of grain she approaches, because the foetus in her is voracious and snatches it. She must not speak to people who are reaping or brewing without first making a ritual gesture of goodwill to cancel the danger. They speak of the foetus 'with jaws agape' snatching food, and explain it by the inevitability of the 'seed within' fighting the 'seed without'.

'The child in the belly . . . is like a witch; it will damage food like witchcraft; beer is spoiled and tastes nasty, food does not grow, the smith's iron is not easily worked, the milk is not good.'

Even the father is endangered at war or in the hunt by his wife's pregnancy (Wilson, pp. 138-9).

Levy-Bruhl noted that menstrual blood and miscarriage sometimes attract the same kind of belief. The Maoris regard menstrual blood as a sort of human being *manqué*. If the blood had not flowed it would have become a person, so it has the impossible status of a dead person that has never lived. He quoted a common belief that a foetus born prematurely has a malevolent spirit, dangerous to the living (pp. 390-6). Levy-Bruhl did not generalise that danger lies in marginal states, but Van Gennep had more sociological insight. He saw society as a house with rooms and corridors in which passage from one to another is dangerous. Danger lies in transitional states, simply because transition is neither one state nor the next, it is undefinable. The person who must pass from one to another is himself in danger and emanates danger to others. The danger is controlled by ritual which precisely separates him from his old status, segregates him for a time and then publicly declares his entry to his new status. Not only is transition itself dangerous, but also the rituals of segregation are the most dangerous phase of the rites. So often do we read that boys die in initiation ceremonies, or that their sisters and mothers are told to fear for their safety, or that they used in the old days to die from hardship or fright, or by supernatural punishment for their misdeeds. Then somewhat tamely come the accounts of the actual ceremonies which are so safe that the threats of danger sound like a hoax (Vansina, 1955). But we can be sure that the trumped up dangers express something important about marginality. To say that the boys risk their lives says precisely that to go out of the formal structure and to enter the margins is to be exposed to power that is enough to kill them or make their manhood. The theme of death and rebirth, of course, has other symbolic functions: the initiates die to their old life and are reborn to the new. The whole repertoire of ideas concerning pollution and purification are used to mark the gravity of the event and the power of ritual to remake a man—this is straightforward.

During the marginal period which separates ritual dying and ritual rebirth, the novices in initiation are temporarily outcast. For the duration of the rite they have no place in society. Sometimes they actually go to live far away outside it. Sometimes they live near enough for unplanned contacts to take place between full social beings and the outcasts. Then we find them behaving like dangerous criminal characters. They are licensed to waylay, steal, rape. This behaviour is even enjoined on them. To behave anti-socially is the proper expression of their marginal condition (Webster, 1908, chapter III). To have been in the margins is to have been in contact with danger, to have been at a source of power. It is consistent with the ideas about form and formlessness to treat initiands coming out of seclusion as if they were themselves charged with power, hot, dangerous, requiring insulation and a time for cooling down. Dirt, obscenity and lawlessness are as relevant symbolically to the rites of seclusion as other ritual expressions of their condition. They are not to be blamed for misconduct any more than the foetus in the womb for its spite and greed.

It seems that if a person has no place in the social system and is therefore a marginal being, all precaution against danger must come from others. He cannot help his abnormal situation. This is roughly how we ourselves regard marginal people in a secular, not a ritual context. Social workers in our society, concerned with the after-care of ex-prisoners, report a difficulty of resettling them in steady jobs, a difficulty which comes from the attitude of society at large. A man who has spent any time 'inside' is put permanently 'outside' the ordinary social system. With no rite of aggregation which can definitively assign him to a new position he remains in the margins, with other people who are similar credited with unreliability, unteachability, and all the wrong social attitudes. The same goes for persons who have entered institutions for the treatment of mental disease. So long as they stay at home their peculiar behaviour is accepted. Once they have been formally classified as abnormal, the very same behaviour is counted intolerable. A report on a Canadian project in 1951 to change the attitude to mental ill-health suggests that there is a threshold of tolerance marked by entry to a mental hospital. If a person has never moved out of society into this marginal state, any of his eccentricities are comfortably tolerated by his neighbours. Behaviour which a psychologist would class at once as pathological is commonly dismissed as 'Just a quirk', or 'He'll get over it', or 'It takes all sorts to make a world'. But once a patient is admitted to a mental hospital, tolerance is withdrawn. Behaviour which was formerly judged to be so normal that the psychologist's suggestions raised strong hostility, was now judged to be abnormal (quoted in Cumming). So mental health workers find exactly the same problems in rehabilitating their discharged patients as do the prisoners' aid societies. The fact that these common assumptions about ex-prisoners and lunatics are self-validating is not relevant here. It is more interesting to know that marginal status produces the same reactions the world over, and that these are deliberately represented in marginal rites.

To plot a map of the powers and dangers in a primitive universe, we need to underline the interplay of ideas of form and formlessness. So many ideas about power are based on an idea of society as a series of forms contrasted with surrounding non-form. There is power in the forms and other power in the inarticulate area, margins, confused lines, and beyond the external boundaries. If pollution is a particular class of danger, to see where it belongs in the universe of dangers we need an inventory of all the possible sources of power. In a primitive culture the physical agency of misfortune is not so significant as the personal intervention to which it can be traced. The effects are the same the world over: drought is drought, hunger is hunger; epidemic, child labour, infirmity—most of the experiences are held in common. But each culture knows a distinctive set of laws governing the way these disasters fall. The main links between persons and misfortunes are personal links. So our inventory of powers must proceed by classifying all kinds of personal intervention in the fortunes of others.

The spiritual powers which human action can unleash can roughly be divided into two classes—internal and external. The first reside within the psyche of the agent—such as evil eye, witchcraft, gifts of vision or prophecy. The second are external symbols on which the agent must consciously work: spells, blessings, curses, charms and formulas and invocations. These powers require actions by which spiritual power is discharged.

This distinction between internal and external sources of power is often correlated with another distinction, between uncontrolled and controlled power. According to widespread beliefs, the internal psychic powers are not necessarily triggered off by the intention of the agent. He may be quite unaware that he possesses them or that they are active. These beliefs vary from place to place. For example, Joan of Arc did not know when her voices would speak to her, could not summon them at will, was often startled by what they said and by the train of events which her obedience to them started. The Azande believe that a witch does not necessarily know that his witch-craft is at work, yet if he is warned, he can exert some control to check its action.

By contrast, the magician cannot utter a spell by mistake; specific intention is a condition of the result. A father's curse usually needs to be pronounced to have effect.

Where does pollution come in the contrast between uncontrolled and controlled power, between psyche and symbol? As I see it, pollution is a source of danger altogether in a different class: the distinctions of voluntary, involuntary, internal, external, are not relevant. It must be identified in a different way.

First to continue with the inventory of spiritual powers, there is another classification according to the social position of those endangering and endangered. Some powers are exerted on behalf of the social structure; they protect society from malefactors against whom their danger is directed. Their use must be approved by all good men. Other powers are supposed to be a danger to society and their use is disapproved; those who use them are malefactors, their victims are innocent and all good men would try to hound them down—these are witches and sorcerers. This is the old distinction between white and black magic.

Are these two classifications completely unconnected? Here I tentatively suggest a correlation: where the social system explicitly recognises positions of authority, those holding such positions are endowed with explicit spiritual power, controlled, conscious, external and approved—powers to bless or curse. Where the social system requires people to hold dangerously ambiguous roles, these persons are credited with uncontrolled, unconscious, dangerous, disapproved powers—such as witchcraft and evil eye.

In other words, where the social system is well-articulated, I look for articulate powers vested in the points of authority; where the social system is ill-articulated, I look for inarticulate powers vested in those who are a source of disorder. I am suggesting that the contrast between form and surrounding non-form accounts for the distribution of symbolic and psychic powers: external symbolism upholds the explicit social structure and internal, unformed psychic powers threaten it from the non-structure.

This correlation is admittedly difficult to establish. For one thing it is difficult to be precise about the explicit social structure. Certainly people carry round with them a consciousness of social structure. They curb their actions in accordance with the symmetries and hierarchies they see therein, and strive continually to impress their view of the relevant bit of structure on other actors in their scene. This social consciousness has been so well demonstrated by Goffman that there should be no need to labour the point further here. There are no items of clothing or of food or of other practical use which we do not seize upon as theatrical props to dramatise the way we want to present our roles and the scene we are playing in. Everything we do is significant, nothing is without its conscious symbolic load. Moreover, nothing is lost on the audience. Goffman uses dramatic structure, with its division of players and audience, stage and back-stage, to provide a frame for his analysis of everyday situations. Another merit of the analogy with theatre is that a dramatic structure exists within temporal divisions. It has a beginning, climax and end. For this reason Turner found it useful to introduce the idea of social drama to describe clusters of behaviour which everyone recognises as forming discrete temporal units (1957). I am sure that sociologists have not finished with the idea of drama as an image of social structure but for my purpose it may be enough to say that by social structure I am not usually referring to a total structure which embraces the whole of society continually and comprehensively. I refer to particular situations in which individual actors are aware of a greater or smaller range of inclusiveness. In these situations they behave as if moving in patterned positions in relation to others, and as if choosing between possible patterns of relations. Their sense of form makes demands on their behaviour, governs their assessment of their desires, permits some and over-rides others.

Any local, personal view of the whole social system will not necessarily coincide with that of the sociologist. Sometimes in what follows, when I speak of social structure, I will be referring to the main outlines, lineages and the hierarchy of descent groups, or chiefdoms and the ranking of districts, relations between royalty and commoners. Sometimes I will be talking about little sub-structures, themselves chinese-box-like, containing others which fill in the bare bones of the main structure. It seems that individuals are aware in appropriate contexts of all these structures and aware of their relative importance. They do not all have the same idea of what particular level of structure is relevant at a given moment; they know there is a problem of communication to be overcome if there can be society at all. By ceremony, speech and gesture they make a constant effort to express and to agree on a view of what the relevant social structure is like. And all the attribution of dangers and powers is part of this effort to communicate and thus to create social forms.

The idea that there may be a correlation between explicit authority and controlled spiritual power was first suggested to me by Leach's article in *Rethinking Anthropology*. In developing the idea I have taken a somewhat different direction. Controlled power to harm, he suggests, is often vested in explicit key points in the authority system, and contrasted with the unintentional power to harm supposed to lurk in the less explicit, weakly articulated areas of the same society. He was mainly concerned with the contrast of two kinds of spiritual power used in parallel contrasting social situations. He presented some societies as sets of internally structured systems interacting with one another. Living within one such system people are explicitly conscious of its structure. Its key points are supported by beliefs in controlled forms of power attached to controlling positions. For instance, Chiefs among the Nyakusa can attack their foes by a kind of sorcery which sends invisible pythons after them. Among the patrilineal Tallensi, a man's father has a correspondingly controlled right of access to ancestral power against him, and among the matrilineal Trobrianders the maternal uncle is thought to support his authority with consciously controlled spells and charms. It is as if the positions of authority were wired up with switches which can be operated by those who reach the right places in order to provide power for the system as a whole.

This can be argued along familiar Durkheimian lines. Religious beliefs express society's awareness of itself; the social structure is credited with punitive powers which maintain it in being. This is quite straightforward. But I would like to suggest that those holding office in the explicit part of the structure tend to be credited with consciously controlled powers, in contrast with those whose role is less explicit and who tend to be credited with unconscious, uncontrollable powers, menacing those in better defined positions. Leach's first example is the

Kachin wife. Linking two power groups, her husband's and her brother's, she holds an interstructural role and she is thought of as the unconscious, involuntary agent of witchcraft. Similarly, the father in the matrilineal Trobrianders and Ashanti, and the mother's brother in patrilineal Tikopia and Taleland, is credited with being an involuntary source of danger. These people are none of them without a proper niche in the total society. But from the perspective of one internal sub-system to which they do not belong, but in which they must operate, they are intruders. They are not suspect in their own system and may be wielding the intentional kind of powers on its behalf. It is possible that their involuntary power to do harm may never be activated. It may lie dormant as they live their life peacefully in the corner of the sub-system which is their proper place, and yet in which they are intruders. But this role is in practice difficult to play coolly. If anything goes wrong, if they feel resentment or grief, then their double loyalties and their ambiguous status in the structure where they are concerned makes them appear as a danger to those belonging fully in it. It is the existence of an angry person in an interstitial position which is dangerous, and this has nothing to do with the particular intentions of the person.

In these cases the articulate, conscious points in the social structure are armed with articulate, conscious powers to protect the system; the inarticulate, unstructured areas emanate unconscious powers which provoke others to demand that ambiguity be reduced. When such unhappy or angry interstitial persons are accused of witchcraft it is like a warning to bring their rebellious feelings into line with their correct situation. If this were found to hold good more generally, then witchcraft, defined as an alleged psychic force, could also be defined structurally. It would be the anti-social psychic power with which persons in relatively unstructured areas of society are credited, the accusation being a means of exerting control where practical forms of control are difficult. Witchcraft, then, is found in the non-structure. Witches are social equivalents of beetles and spiders who live in the cracks of the walls and wainscoting. They attract the fears and dislikes which other ambiguities and contradictions attract in other thought structures, and the kind of powers attributed to them symbolise their ambiguous, inarticulate status.

Powers and Dangers

Pondering on this line of thought, we can distinguish different types of social inarticulateness. So far we have only considered witches who have a well-defined position in one sub-system and an ambiguous one in another, in which they none the less have duties. They are legitimate intruders. Of these Joan of Arc can be taken as a splendid prototype: a peasant at court, a woman in armour, an outsider in the councils of war; the accusation that she was a witch puts her fully in this category.

But witchcraft is often supposed to operate in another kind of ambiguous social relation. The best example comes from the witchcraft beliefs of the Azande. The formal structure of their society was pivoted on princes, their courts, tribunals and armies, in a clear cut hierarchy down to princes' deputies, through local governors, to heads of families. The political system afforded an organised set of fields for competition, so that commoners did not find themselves in competition with nobles, nor poor against rich, nor sons against fathers, nor women against men. Only in those areas of society which were left unstructured by the political system did men accuse each other of witchcraft. A man who had defeated a close rival in competition for office might accuse the other of bewitching him in jealousy, and co-wives might accuse one another of witchcraft. Azande witches were thought to be dangerous without knowing it; their witchcraft was made active simply by their feelings of resentment or grudge. The accusation attempted to regulate the situation by vindicating one and condemning the other rival. Princes were supposed not to be witches, but they accused one another of sorcery, thus conforming to the pattern I am seeking to establish.

Another type of unconscious power to harm emanating from inarticulate areas of the social system is illustrated by the Mandari, whose land-owning clans build up their strength by adopting clients. These unfortunates have, for one reason or another, lost their claim to their own territory and have come to a foreign territory to ask for protection and security. They are landless, inferior, dependent on their patron who is a member of a land-owning group. But they are not completely dependent. To some real extent the patron's influence and status depend on his loyal following of clients. Clients who become too numerous and bold can threaten their patron's lineage. The explicit structure of society is based on land-holding clans. By these people clients are held likely to be witches. Their witchcraft emanates from jealousy of their patrons and works involuntarily. A witch cannot control himself, it is his nature to be angry and harm emanates from him. Not all clients are witches, but hereditary lines of witches are recognised and feared. Here are people living in the interstices of the power structure, felt to be a threat to those with better defined status. Since they are credited with dangerous, uncontrollable powers, an excuse is given for supressing them. They can be charged with witchcraft and violently despatched without formality or delay. In one case the patron's family merely made ready a big fire, called in the suspect witch to share a meal of roast pig, and forthwith bound him and put him on the fire. Thus the formal structure of land-holding lineages was asserted against the relatively fluid field in which landless clients touted for patronage.

Jews in English society are something like Mandari clients. Belief in their sinister but undefinable advantages in commerce justifies discrimination against them—whereas their real offence is always to have been outside the formal structure of Christendom.

There are probably many more variant types of socially ambiguous or weakly defined statuses to which involuntary witchcraft is attributed. It would be easy to go on piling up examples. Needless to say, I am not concerned with beliefs of a secondary kind or with short-lived ideas which flourish briefly and die away. If the correlation were generally to hold good for the distribution of dominant, persistent forms of spiritual power, it would clarify the nature of pollution. For, as I see it, ritual pollution also arises from the interplay of form and surrounding formlessness. Pollution dangers strike when form has been attacked. Thus we would have a triad of powers controlling fortune and misfortune: first, formal powers wielded by persons representing the formal structure and exercised on behalf of the formal structure: second, formless powers wielded by interstitial persons: third, powers not wielded by any person, but inhering in the structure, which strike against any infraction of form. This three-fold scheme for investigating primitive cosmologies unfortunately comes to grief over exceptions which are too important to brush aside. One big difficulty is that sorcery, which is a form of controlled spiritual power, is in many parts of the world credited to persons who ought, according to my hypothesis, be charged with involuntary witchcraft. Malevolent persons in interstitial positions, anti-social, disapproved, working to harm the innocent, they should not be using conscious, controlled, symbolic power. Furthermore, there are royal chiefs who emanate unconscious, involuntary power to detect disaffection and destroy their enemies—chiefs who according to my hypothesis should be content with explicit, controlled forms of power. So the correlation I have tried to draw does not hold. However, I will not throw it aside until I have looked more closely at the negative cases.

One reason why it is difficult to correlate social structure with type of mystic power is that both elements in the comparison are very complex. It is not always easy to recognise explicit authority. For example, authority among the Lele is very weak, their social system makes a criss-cross of little authorities, none very effective in secular terms. Many of their formal statuses are supported by the spiritual power to curse or bless, which consists in uttering a form of words and spitting. Cursing and blessing are attributes of authority; a father, mother, mother's brother, aunt, pawn owner, village head and so on, can curse. Not any one can reach out for a curse and apply it arbitrarily. A son cannot curse his father, it would not work if he tried. So this pattern conforms to the general rule I am seeking to establish. But, if a person who has a right to curse refrains from formulating his curse, the unspit saliva in his mouth is held to have power to cause harm. Better than harbour a secret grudge, anyone with a just grievance should speak up and demand redress, lest the saliva of his ill-will do harm secretly. In this belief we have both the controlled and uncontrolled spiritual power attributed to the same person in the same circumstances. But as their pattern of authority is so weakly articulated, this is hardly a negative case. On the contrary, it serves to warn us that authority can be a very vulnerable power, easily reduced to nothing. We should be prepared to elaborate the hypothesis to take more account of the varieties of authority.

There are several likenesses between the unspoken curse of the Lele and the witchcraft beliefs of the Mandari. Both are tied to a particular status, both are psychic, internal, involuntary. But the unspoken curse is an approved form of spiritual power, while the witch is disapproved. Where the unspoken curse is revealed as the cause of harm restitution is made to the agent, when witchcraft is revealed the agent is brutally attacked. So the unspoken curse is on the side of authority; its link with cursing makes this clear. But authority is weak in the case of the Lele, strong in the case of the Mandari. This suggests that to test the hypothesis fairly we should display the whole gamut from no formal authority at one end of the scale to strong effective secular authority at the other end. At either extreme I am not prepared to predict the distribution of spiritual powers, because where there is no formal authority the hypothesis does not apply, and where authority is firmly established by secular means it less requires spiritual and symbolic support. Under primitive conditions authority is always likely to be precarious. For this reason we should be ready to take into account the failure of those in office.

First consider the case of the man in a position of authority who abuses the secular powers of his office. If it is clear that he is acting wrongly, out of role, he is not entitled to the spiritual power which is vested in the role. Then there should be scope for some shift in the pattern of beliefs to accommodate his defection. He ought to enter the class of witches, exerting involuntary, unjust powers instead of intentionally controlled powers against wrongdoers. For the official who abuses his office is as illegitimate as an usurper, an incubus, a spanner in the works, a dead weight on the social system. Often we find this predicted shift in the kind of dangerous power he is supposed to wield.

In the Book of Samuel, Saul is presented as a leader whose divinely given powers are abused. When he fails to fill his assigned role and leads his men into disobedience, his charisma leaves him and terrible rages, depression and madness afflict him. So when Saul abuses his office he loses conscious control and becomes a menace even to his friends. With reason no longer in control, the leader becomes an unconscious danger. The image of Saul fits the idea that conscious spiritual power is vested in the explicit structure and uncontrolled unconscious danger vested in the enemies of the structure.

The Lugbara have another and similar way of adjusting their beliefs to abuse of power. They credit their lineage elders with special powers to invoke the ancestors against juniors who do not act in the widest interests of the lineage. Here again we have conscious controlled powers upholding the explicit structure.

But if an elder is thought to be motivated by his own personal, selfish interests, the ancestors neither listen to him nor put their power at his disposal. So here is a man in a position of authority, improperly wielding the powers of office. His legitimacy being in doubt, he must be removed, and to remove him his antagonists accuse him of having become corrupt and emanating witchcraft, a mysterious, perverted power which operates at night (Middleton). The accusation is itself a weapon for clarifying and strengthening the structure. It enables guilt to be pinned on the source of confusion and ambiguity. So these two examples symmetrically develop the idea that conscious power is exerted from the key positions in the structure and a different danger from its dark, obscure areas.

Sorcery is another matter. As a form of harmful power which makes use of spells, words, actions and physical materials, it can only be used consciously and deliberately. On the argument we have been following, sorcery ought to be used by those in control of key positions in the social structure as it is a deliberate, controlled form of spiritual power. But it is not. Sorcery is found in the structural interstices where we have located witchcraft, as well as in the seats of authority. At first glance it seems to cut across the correlation of articulate structure with consciousness. But on closer inspection this distribution of sorcery is consistent with the pattern of authority that goes with sorcery beliefs.

In some societies positions of authority are open to competition. Legitimacy is hard to establish, hard to maintain and always liable to reversal. In such very fluid political systems we would expect a particular type of beliefs in spiritual power. Sorcery is unlike cursing and invocation of ancestors in that it has no built-in device to safeguard against abuse. Lugbara cosmology, for example, is dominated by the idea of the ancestors upholding lineage values; the Israelite cosmology was dominated by the idea of the justice of Jehovah. Both these sources of power contain an assumption that they cannot be deceived or abused. If an incumbent of office misuses his power, spiritual support is withdrawn. By contrast, sorcery is essentially a form of controlled and conscious power that is open to abuse. In the Central African cultures, where sorcery beliefs flourish, this form of spiritual power is developed within the idiom of medicine. It is freely available. Anyone who takes the trouble to acquire sorcery power may use it. In itself it is morally and socially neutral and it contains no principle for safeguarding against abuse. It works *ex opere operato*, equally well whether the intentions of the agent are pure or corrupt. If the idea of spiritual power in the culture is dominated by this medical idiom, the man who abuses his office and the person in the unstructured crevices have the same access to the same kind of spiritual powers as the lineage or village head. It follows that if sorcery is available to anyone who wants to acquire it, then we should suppose that positions of political control are also available, open to competition, and that in such societies there are not very clear distinctions between legitimate authority, abuse of authority and illegitimate rebellion.

The sorcery beliefs of Central Africa, west to east from the Congo to Lake Nyasa, assume that malign spiritual powers of sorcery are generally available. In principle these powers are vested in the heads of matrilineal descent groups and are expected to be used by these men in authority against enemy outsiders. There is a general expectation that the old man may turn his powers against his own followers and kin. and if he is disagreeable or mean, their deaths are likely to be attributed to him. He always risks being dragged down from his little elevation of senior status, degraded, exiled or put to the poison ordeal (Van Wing, p. 359-60, Kopytoff, p. 90). Then another contender will take his official role and try to exercise it more warily. Such beliefs, as I have tried to show in my study of the Lele, correspond to a social system in which authority is weakly defined and has little real sway (1963). Marwick has claimed for similar beliefs among the Cewa that they have a liberating effect, since any young man can plausibly accuse of sorcery a reactionary old incumbent of an office which he himself is qualified to occupy when the senior obstacle has been removed (1952). If sorcery beliefs really serve as instruments for self-promotion they also ensure that the ladder of promotion is short and shaky.

The fact that anyone may lay hands on sorcery power and that it is available for use against, or on behalf of society suggests another cross-classification of spiritual powers. For in Central Africa sorcery is often a necessary adjunct to roles of authority. The mother's brother must be acquainted with sorcery to be able to combat enemy sorcerers and to protect his descendants. It is a double-edged attribute, for if he uses it unwisely he can be ruined. Thus there is always the possibility, even the expectation, that the man in an official position will fail to fill it creditably. The belief acts as a check on the use of secular power. If a leader among the Cewa or Lele becomes unpopular the sorcery beliefs contain an escape clause enabling his dependents to get rid of him. This is how I read the Tsav beliefs of the Tiv, checking as much as validating the eminent lineage elder's authority (Bohannan). So freely available sorcery is a form of spiritual power biased towards failure. This is a cross-classification which puts witchcraft and sorcery in the same bracket. Witchcraft beliefs are also tilted to expect role failure and to deal with it punitively, as we have seen. But witchcraft beliefs expect failure in interstitial roles, while sorcery beliefs expect failure in official roles. The whole scheme in which spiritual powers are correlated with structure becomes more consistent if we contrast those powers which are biased towards failure with powers which are biased towards success.

Teutonic notions of Luck, and some forms of *baraka* and *mana* are success-biased beliefs which parallel sorcery as a failure-biased belief. *Mana* and Islamic *baraka* exude from official positions, regardless of the intention of the incumbent. They are either dangerous powers to strike or benign powers for good. There are chiefs and princes exerting *mana* or *baraka* whose merest contact is worth a blessing and a guarantee of success, and whose personal presence makes the difference between victory and defeat in battle. But these powers are not always so well anchored to the outlines of the social system. Sometimes *baraka* can be a free-floating benign power, working independently of the formal distribution of power and allegiance in society.

If we find such free-lance benign contagion playing an important role in people's beliefs, we can expect either that formal authority is weak or ill-defined or that, for one reason or another, the political structure has been neutralised so that the powers of blessing cannot emanate from its key points.

Dr. Lewis has described an example of an un-sacralised social structure. In Somaliland there is a general division in thought between secular and spiritual power (1963). In secular relations power derives from fighting strength and the Somali are militant and competitive. The political structure is a warrior system where might is right. But in the religious sphere the Somalis are Muslims and hold that fighting within the Muslim community is wrong. These deeply held beliefs de-ritualise the social structure so that Somali do not claim that divine blessings or dangers emanate from its representatives. Religion is represented not by warriors but by men of God. These holy men, religious and legal experts, mediate between men as they mediate between men and God. They are only reluctantly involved in the warrior structure of society. As men of God they are credited with spiritual power. It follows that their blessing (*baraka*) is great in proportion as they withdraw from the secular world and are humble, poor and weak.

If this argument is correct it should apply to other Islamicised peoples whose social organisation is based on violent internal conflict. However the Moroccan Berbers exhibit a similar distribution of spiritual power without the theological justification. Professor Gellner tells me that Berbers have no notion that fighting within the Moslem community is wrong. Moreover it is a common feature of competitive segmentary political systems that the leaders of the aligned forces enjoy less credit for spiritual power than certain persons in the interstices of political alignment. The Somali holy man should be seen as the counterpart of the Tallensi Earth shrine priest and the Nuer Man of the Earth. The paradox of spiritual power vested in the physically weak is explained by social structure rather than by the local doctrine which justifies it. (Fortes and Evans-Pritchard, 1940, p.22).

Baraka in this form is something like witchcraft in reverse. It is a power which does not belong to the formal political structure, but which floats between its segments. As witchcraft accusations are used to reinforce the structure, so do people in the structure try to make use of *baraka*. Like witchcraft and sorcery its existence and strength is proved empirically, *post hoc*. A witch or sorcerer is identified when a misfortune occurs to someone against whom he has a grudge. The misfortune indicates there is witchcraft at work. The known grudge indicates the possible witch. It is his reputation for quarrels which essentially focusses the charge against him. *Baraka* is also identified empirically, *post hoc*. A piece of marvellous good fortune indicates its presence, often quite unexpectedly (Westermarck, I, chapter II). The reputation of a holy man for piety and learning focusses interest on him. Just as the witch's bad name will get worse with every disaster that befalls her neighbours, so the saint's good name will improve with every stroke of good fortune. The snow-ball effect is the same.

The failure-biased powers have a negative feed-back. If anyone potentially possessing them tries to get above himself, the accusation cuts him down to size. The fear of accusation works like a thermostat on everyone in advance of actual quarrels. It is a control device. But the success-biased powers have the possibility of a positive feed-back. They could build up and up indefinitely to an explosion. As witchcraft has been called institutionalised jealousy, so *baraka* can work as institutionalised admiration. For this reason it is self-validating when it works in a freely competitive system. It is on the side of the big battalions. Empirically tested by success, it attracts adherents and so earns more success. 'People in fact become possessors of *baraka* by being treated as possessors of it.' (Gellner 1962).

I should make it clear that I do not believe that *baraka* is always available to competing elements in tribal social systems. It is an idea about power which varies in different political conditions. In an authoritative system it can emanate from the holders of authority and validate their established status, to the discomfiture of their foes. But it also has the potentiality of disrupting ideas about authority and about right and wrong, since its only proof lies in its success. The possessor of *baraka* is not subject to the same moral restraints as other persons (Westermarck, I, p. 198). The same applies to *Mana* and Luck. They can be on the side of established authority or on the side of opportunism. Raymond Firth came to the conclusion that at least in Tikopia, *Mana* means success (1940). Tikopian *Mana* expresses the authority of hereditary chiefs. Firth reflected on whether the dynasty would be endangered if the chief's reign were not a fortunate one, and concluded (correctly as it happened) that the chiefship would be strong enough to ride such a storm. One of the great advantages of doing sociology in a teacup is to be able to discern calmly what would be confusing in a larger scene. But it is a drawback not to be able to observe any real storms and upheavals. In a sense all colonial anthropology takes place in a teacup. If *mana* means success it is an apt concept for political opportunism. The artificial conditions of colonial peace may have disguised this potential for conflict and rebellion which the success-biased powers imply. Anthropology has often been weak in political analysis. The equivalent of a paper constitution without any dust or conflict or serious estimate of the balance of forces is sometimes offered in lieu of an analysis of a political system. This must necessarily obscure interpretation. It may be helpful to turn to a pre-colonial example.

Luck, for our Teutonic ancestors, like the opportunist or freelance forms of *mana* and *baraka*, also seems to have operated freely in a competitive political structure, fluid, with little in the way of hereditary power. Such beliefs can follow swift changes in the lines of allegiance, and change judgments of right and wrong.

I have tried to push as far as possible the parallel between these success-biased powers and witchcraft and sorcery, both failure-biased and both capable of operating independently of the distribution of authority. Another parallel with witchcraft is in the involuntary nature of these success forces. A man discovers he has *baraka* because of its effects. Many men may be pious and live outside the warrior system, but not many have great *baraka*. *Mana* too may be exerted quite unconsciously, even by the anthropologist, as Raymond Firth wryly recounts when a magnificent haul of fish was attributed to his *mana*. The Sagas of the Norsemen are full of crises resolved when a man suddenly discovers his Luck or finds that his Luck has deserted him (Grönbech, Vol. I, ch. 4).

Another characteristic of success power is that it is often contagious. It is transmitted materially. Anything which has been in contact with *baraka* may get *baraka*. Luck was also transmitted partly in heirlooms and treasures. If these changed hands, Luck changed hands too. In this respect these powers are like pollution, which transmits danger by contact. However, the potentially haphazard and disruptive effects of these success powers contrasts with pollution, austerely committed to support the outlines of the existing social system.

To sum up, beliefs which attribute spiritual power to individuals are never neutral or free of the dominant patterns of social structure. If some beliefs seem to attribute free-floating spiritual powers in a haphazard manner, closer inspection shows consistency. The only circumstances in which spiritual powers seem to flourish independently of the formal social system are when the system itself is exceptionally devoid of formal structure, when legitimate authority is always under challenge or when the rival segments of an acephalous political system resort to mediation. Then the main contenders for political power have to court for their side the holders of free-floating spiritual power. Thus it is beyond doubt that the social system is thought of as quick with creative and sustaining powers.

Now is the time to identify pollution. Granted that all spiritual powers are part of the social system. They express it and provide institutions for manipulating it. This means that the power in the universe is ultimately hitched to society, since so many changes of fortune are set off by persons in one kind of social position or another. But there are other dangers to be reckoned with, which persons may set off knowingly or unknowingly, which are not part of the psyche and which are not to be bought or learned by initiation and training. These are pollution powers which inhere in the structure of ideas itself and which punish a symbolic breaking of that which should be joined or joining of that which should be separate. It follows from this that pollution is a type of danger which is not likely to occur except where the lines of structure, cosmic or social, are clearly defined.

A polluting person is always in the wrong. He has developed some wrong condition or simply crossed some line which should not have been crossed and this displacement unleashes danger for someone. Bringing pollution, unlike sorcery and witchcraft, is a capacity which men share with animals, for pollution is not always set off by humans. Pollution can be committed intentionally, but intention is irrelevant to its effect—it is more likely to happen inadvertently.

This is as near as I can get to defining a particular class of dangers which are not powers vested in humans, but which can be released by human action. The power which presents a danger for careless humans is very evidently a power inhering in the structure of ideas, a power by which the structure is expected to protect itself.

7

External Boundaries

THE IDEA OF SOCIETY is a powerful image. It is potent in its own right to control or to stir men to action. This image has form; it has external boundaries, margins, internal structure. Its outlines contain power to reward conformity and repulse attack. There is energy in its margins and unstructured areas. For symbols of society any human experience of structures, margins or boundaries is ready to hand.

Van Gennep shows how thresholds symbolise beginnings of new statuses. Why does the bridegroom carry his bride over the lintel? Because the step, the beam and the door posts make a frame which is the necessary everyday condition of entering a house. The homely experience of going through a door is able to express so many kinds of entrance. So also are cross roads and arches, new seasons, new clothes and the rest. No experience is too lowly to be taken up in ritual and given a lofty meaning. The more personal and intimate the source of ritual symbolism, the more telling its message. The more the symbol is drawn from the common fund of human experience, the more wide and certain its reception.

The structure of living organisms is better able to reflect complex social forms than door posts and lintels. So we find that the rituals of sacrifice specify what kind of animal shall be used, young or old, male, female or neutered, and that these rules signify various aspects of the situation which calls for sacrifice. The way the animal is to be slaughtered is also laid down. The Dinka cut the beast longitudinally through the sexual organs if the sacrifice is intended to undo an incest; in half across the middle for celebrating a truce; they suffocate it for some occasions and trample it to death for others. Even more direct is the symbolism worked upon the human body. The body is a model which can stand for any bounded system. Its boundaries can represent any boundaries which are threatened or precarious. The body is a complex structure. The functions of its different parts and their relation afford a source of symbols for other complex structures. We cannot possibly interpret rituals concerning excreta, breast milk, saliva and the rest unless we are prepared to see in the body a symbol of society, and to see the powers and dangers credited to social structure reproduced in small on the human body.

It is easy to see that the body of a sacrificial ox is being used as a diagram of a social situation. But when we try to interpret rituals of the human body in the same way the psychological tradition turns its face away from society, back towards the individual. Public rituals may express public concerns when they use inanimate door posts or animal sacrifices: but public rituals enacted on the human body are taken to express personal and private concerns. There is no possible justification for this shift of interpretation just because the rituals work upon human flesh. As far as I know the case has never been methodically stated. Its protagonists merely proceed from unchallenged assumptions, which arise from the strong similarity between certain ritual forms and the behaviour of psychopathic individuals. The assumption is that in some sense primitive cultures correspond to infantile stages in the development of the human psyche. Consequently such rites are interpreted as if they express the same preoccupations which fill the mind of psychopaths or infants.

Let me take two modern attempts to use primitive cultures to buttress psychological insights. Both stem from a long line of similar discussions, and both are misleading because the relation between culture and individual psyche are not made clear.

Bettelheim's *Symbolic Wounds* is mainly an interpretation of circumcision and initiation rites. The author tries to use the set rituals of Australians and Africans to throw light on psychological phenomena. He is particularly concerned to show that psychoanalysts have over-emphasised girls' envy of the male sex and overlooked the importance of boys' envy of the female sex. The idea came to him originally in studying groups of schizophrenic children approaching adolescence. It seems very likely that the idea is sound and important. I am not at all claiming to criticise his insight into schizophrenia. But when he argues that rituals which are explicitly designed to produce genital bleeding in males are intended to express male envy of female reproductive processes, the anthropologist should protest that this is an inadequate interpretation of a public rite. It is inadequate because it is merely descriptive. What is being carved in human flesh is an image of society. And in the moiety- and section-divided tribes he cites, the Murngin and Arunta, it seems more likely that the public rites are concerned to create a symbol of the symmetry of the two halves of society.

The other book is *Life against Death*, in which Brown outlines an explicit comparison between the culture of 'archaic man' and our own culture, in terms of the infantile and neurotic fantasies which they seem to express. Their common assumptions about primitive culture derive from Roheim (1925): primitive culture is autoplastic, ours is alloplastic. The primitive seeks to achieve his desires by self-manipulation, performing surgical rites upon his own body to produce fertility in nature, subordination in women or hunting success. In modern culture we seek to achieve our desires by operating directly on the external environment, with the impressive technical results that are the most obvious distinction between the two types of cultures. Bettelheim adopts this summing up of the difference between the ritual and the technical bias in civilisation. But he supposes that the primitive culture is produced by inadequate, immature personalities, and even that the psychological shortcomings of the savage accounts for his feeble technical achievements:

'If preliterate peoples had personality structures as complex as those of modern man, if their defences were as elaborate and their consciences as refined and demanding; if the dynamic interplay between ego, superego and id were as intricate and if their ego's were as well adapted to meet and change external reality —they would have developed societies equally complex, though probably different. Their societies have, however, remained small and relatively ineffective in coping with the external environment. It may be that one of the reasons for this is their tendency to try to solve problems by autoplastic rather than alloplastic manipulation.' (p. 87)

Let us assert again, as many anthropologists have before, that there are no grounds for supposing that primitive culture as such is the product of a primitive type of individual whose personality resembles that of infants or neurotics. And let us challenge the psychologists to express the syllogisms on which such a hypothesis might rest. Underlying the whole argument is the assumption that the problems which rituals are intended to solve are personal psychological problems. Bettelheim actually goes on to compare the primitive ritualist with the child who hits his own head when frustrated. This assumption underlies his whole book.

Brown makes the same assumption, but his reasoning is more subtle. He does not suppose that the culture's primitive condition is caused by individual personal traits: he allows very properly for the effect of cultural conditioning on the individual personality. But he proceeds to consider the total culture as if it, in its totality, could be compared to an infant or a retarded adult. The primitive culture resorts to bodily magic to achieve its desires. It is in a stage of cultural evolution comparable to that of infantile anal eroticism. Starting from the maxim:

> 'Infantile sexuality is autoplastic compensation for the loss of the Other; sublimation is alloplastic compensation for loss of Self.' (p. 170)

he goes on to argue that 'archaic' culture is directed to the same ends as infantile sexuality, that is escape from the hard realities of loss, separation and death. Epigrams are, by their nature, obscure. This is another approach to primitive culture which I would like to see fully spelt out. Brown develops the theme only briefly, as follows:

> 'Archaic man is preoccupied with the castration complex, the incest taboo and the desexualisation of the penis, that is, the transference of the genital impulses into that aim-inhibited libido which sustains the kinship systems in which archaic life is embedded. The low degree of sublimation, corresponding to the low level of technology, means by our previous definitions, a weaker ego, an ego which has not yet come to terms (by negation) with its own pregenital impulses. The result is that all the fantastic wishes of infantile narcissism express themselves in unsublimated form so that archaic man retains the magic body of infancy.' (p. 298-9)

These fantasies suppose that the body itself could fulfil the infant's wish for unending, self-replenishing enjoyment. They are a flight from reality, a refusal to face loss, separation and death. The ego develops by sublimating these fantasies. It mortifies the body, denies the magic of excrement and to that extent faces reality. But sublimation substitutes another set of unreal aims and ends by providing the self with another kind of false escape from loss, separation and death. This is how I understand the argument to run. The more material that an elaborate technology imposes between ourselves and the satisfaction of our infantile desires, the more busily has sublimation been at work. But the converse seems questionable. Can we argue that the less the material basis of civilisation is developed, the less sublimation has been at work? What precise analogy with infantile fantasy can be valid for a primitive culture based on a primitive technology? How does a low level of technology imply 'an ego which has not yet come to terms (by negation) with its own pregenital impulses'? In what sense is one culture more sublimated than another?

These are obviously technical questions in which the anthropologist cannot engage. But on two points the anthropologist has something to say. One is the question of whether primitive cultures really can be said to revel in excremental magic. The answer to this is surely No. The other is whether primitive cultures seem to be seeking an escape from reality. Do they really use their magic, excremental or other, to compensate for loss of success in external fields of endeavour? Again the answer is No.

To take the matter of excremental magic first. The information is distorted, first as to the relative emphasis on bodily as distinct from other symbolic themes, and second as to the positive or negative attitudes to bodily refuse seen in primitive ritual.

To take up the latter point first: the use of excrement and other bodily exuviae in primitive cultures is usually inconsistent with the themes of infantile erotic fantasy. So far from excrement, etc., being treated as a source of gratification, its use tends to be condemned. So far from being thought of as an instrument of desire, the power residing in the margins of the body is more often to be avoided. There are two main reasons why casual reading in ethnography gives the wrong impression. The first is an informant's bias and the second an observer's bias.

Sorcerers are supposed to use bodily refuse in pursuing their nefarious desires. Certainly in this sense excremental magic ministers to its user's desires, but information about sorcery is usually given from the alleged victim's point of view. Vivid accounts of the *materia medica* of sorcery can always be had from supposed victims. But recipe books of charms dictated by confessed sorcerers are rarer. It is one thing to suspect that others are using bodily refuse unlawfully against oneself, but this does not mean that informants think of these materials as available for their own use. So a kind of optical delusion makes what often belongs on the negative side of the balance sheet appear on the positive side.

The observer's bias also exaggerates the extent to which primitive cultures make positive magical use of bodily relicts. For various reasons best known to psychologists, any reference to excremental magic seems to leap to the reader's eye and absorb attention. Thus a second distortion is introduced. The full richness and range of symbolism tends to be overlooked, or assimilated to a few scatologic principles. Take as an illustration of this bias Brown's own discussion of the Trickster Myth of the Winnebago Indians which we mentioned in Chapter 3. Anal topics occur only two or three times in the course of the long series of Trickster's adventures. I quoted one of these occasions, where Trickster tried to treat his anus as a separate person. Brown's impression of the myth is so different that at first I mistakenly thought he had gone back in erudite fashion to a more primary source than Radin's when he said that:

'The Trickster of primitive mythologies is surrounded by unsublimated and undisguised anality.'

According to Brown the Winnebago Trickster, who is also a great culture hero, 'can create the world by a filthy trick out of excrement, mud, clay'. He cites as example an episode in the myth in which Trickster defies a warning not to eat some bulb which fills his belly with wind, each eruption of which lifts him higher and higher. He calls the humans to hold him down, but in thanks for their attempt to help him in a last final eruption he scatters them all far and wide. Search the story as told by Radin in vain for any sign that Trickster's defaecation is creative in any way. It is rather destructive. Search Radin's glossary and introduction and learn that Trickster did not create the world and is not in any sense a culture hero. Radin considers the quoted episode to have an altogether negative moral, and one consistent with the theme of Trickster's gradual development as a social being. So much for the bias which reads too much excremental magic into primitive cultures.

The next point concerning cultural parallels with anal eroticism is to ask in what sense any primitive cultures are in flight from the realities of separation and loss. Do they try to ignore the unity of death and life? On the contrary, my impression is that those rituals which most explicitly credit corrupt matter with power are making the greatest effort to affirm the physical fullness of reality. So far from using bodily magic as an escape, cultures which frankly develop bodily symbolism may be seen to use it to confront experience with its inevitable pains and losses. By such themes they face the great paradoxes of existence, as I shall show in the last chapter. Here I only touch on the subject briefly because it bears on the parallel with infantile psychology as follows: insofar as ethnography supports the idea that primitive cultures treat dirt as a creative power it contradicts the idea that these cultural themes can be compared with the fantasies of infantile sexuality.

To correct the two distortions of evidence to which this subject is prone we should classify carefully the contexts in which body dirt is thought of as powerful. It may be used ritually for good, in the hands of those vested with power to bless. Blood, in Hebrew religion, was regarded as the source of life, and not to be touched except in the sacred conditions of sacrifice. Sometimes the spittle of persons in key positions is thought effective to bless. Sometimes the cadaver of the last incumbent yields up material for anointing his royal successor. For example, the decayed corpse of the last Lovedu queen in the Drakensberg mountains is used to concoct unguents which enable the current queen to control the weather (Krige, pp. 273-4). These examples can be multiplied. They repeat the analysis in the previous chapter of the powers attributed to the social or religious structure for its own defence. The same goes for body dirt as ritual instrument of harm. It may be credited to the incumbents of key positions for defending the structure, or to sorcerers abusing their positions in the structure, or to outsiders hurling bits of bone and other stuff at weak points in the structure.

But now we are ready to broach the central question. Why should bodily refuse be a symbol of danger and of power? Why should sorcerers be thought to qualify for initiation by shedding blood or committing incest or anthropophagy? Why, when initiated, should their art consist largely of manipulating powers thought to inhere in the margins of the human body? Why should bodily margins be thought to be specially invested with power and danger?

First, we can rule out the idea that public rituals express common infantile fantasies. These erotic desires which it is said to be the infant's dream to satisfy within the body's bounds are presumably common to the human race. Consequently body symbolism is part of the common stock of symbols, deeply emotive because of the individual's experience. But rituals draw on this common stock of symbols selectively. Some develop here, others there. Psychological explanations cannot of their nature account for what is culturally distinctive.

Second, all margins are dangerous. If they are pulled this way or that the shape of fundamental experience is altered. Any structure of ideas is vulnerable at its margins. We should expect the orifices of the body to symbolise its specially vulnerable points. Matter issuing from them is marginal stuff of the most obvious kind. Spittle, blood, milk, urine, faeces or tears by simply issuing forth have traversed the boundary of the body. So also have bodily parings, skin, nail, hair clippings and sweat. The mistake is to treat bodily margins in isolation from all other margins. There is no reason to assume any primacy for the individual's attitude to his own bodily and emotional experience, any more than for his cultural and social experience. This is the clue which explains the unevenness with which different aspects of the body are treated in the rituals of the world. In some, menstrual pollution is feared as a lethal danger; in others not at all (see Chapter 9). In some, death pollution is a daily preoccupation; in others not at all. In some, excreta is dangerous, in others it is only a joke. In India cooked food and saliva are pollution-prone, but Bushmen collect melon seeds from their mouths for later roasting and eating (Marshall Thomas, p. 44).

Each culture has its own special risks and problems. To which particular bodily margins its beliefs attribute power depends on what situation the body is mirroring. It seems that our deepest fears and desires take expression with a kind of witty aptness. To understand body pollution we should try to argue back from the known dangers of society to the known selection of bodily themes and try to recognise what appositeness is there.

In pursuing a last-ditch reduction of all behaviour to the personal preoccupations of individuals with their own bodies the psychologists are merely sticking to their last.

'The derisive remark was once made against psychoanalysis that the unconscious sees a penis in every convex object and a vagina or anus in every concave one. I find that this sentence well characterises the facts.'

(Ferenczi, *Sex in Psychoanalysis*, p. 227, quoted by Brown)

It is the duty of every craftsman to stick to his last. The sociologists have the duty of meeting one kind of reductionism with their own. Just as it is true that everything symbolises the body, so it is equally true (and all the more so for that reason) that the body symbolises everything else. Out of this symbolism, which in fold upon fold of interior meaning leads back to the experience of the self with its body, the sociologist is justified in trying to work in the other direction to draw out some layers of insight about the self's experience in society.

If anal eroticism is expressed at the cultural level we are not entitled to expect a population of anal erotics. We must look around for whatever it is that has made appropriate any cultural analogy with anal eroticism. The procedure in a modest way is like Freud's analysis of jokes. Trying to find a connection between the verbal form and the amusement derived from it he laboriously reduced joke interpretation to a few general rules. No comedian script-writer could use the rules for inventing jokes, but they help us to see some connections between laughter, the unconscious, and the structure of stories. The analogy is fair for pollution is like an inverted form of humour. It is not a joke for it does not amuse. But the structure of its symbolism uses comparison and double meaning like the structure of a joke.

Four kinds of social pollution seem worth distinguishing. The first is danger pressing on external boundaries; the second, danger from transgressing the internal lines of the system; the third, danger in the margins of the lines. The fourth is danger from internal contradiction, when some of the basic postulates are denied by other basic postulates, so that at certain points the system seems to be at war with itself. In this chapter I show how the symbolism of the body's boundaries is used in this kind of unfunny wit to express danger to community boundaries.

The ritual life of the Coorgs (Srinivas) gives the impression of a people obsessed by the fear of dangerous impurities entering their system. They treat the body as if it were a beleaguered town, every ingress and exit guarded for spies and traitors. Anything issuing from the body is never to be re-admitted, but strictly avoided. The most dangerous pollution is for anything which has once emerged gaining re-entry. A little myth, trivial by other standards, justifies so much of their behaviour and system of thought that the ethnographer has to refer to it three or four times. A Goddess in every trial of strength or cunning defeated her two brothers. Since future precedence depended on the outcome of these contests, they decided to defeat her by a ruse. She was tricked into taking out of her mouth the betel that she was chewing to see if it was redder than theirs and into popping it back again. Once she had realised she had eaten something which had once been in her own mouth and was therefore defiled by saliva, though she wept and bewailed she accepted the full justice of her downfall. The mistake cancelled all her previous victories, and her brothers' eternal precedence over her was established as of right.

The Coorgs have a place within the system of Hindoo castes. There is good reason to regard them as not exceptional or aberrant in Hindoo India (Dumont and Pocock). Therefore they conceive status in terms of purity and impurity as these ideas are applied throughout the regime of castes. The lowest castes are the most impure and it is they whose humble services enable the higher castes to be free of bodily impurities. They wash clothes, cut hair, dress corpses and so on. The whole system represents a body in which by the division of labour the head does the thinking and praying and the most despised parts carry away waste matter. Each sub-caste community in a local region is conscious of its relative standing in the scale of purity. Seen from ego's position the system of caste purity is structured upwards. Those above him are more pure. All the positions below him, be they ever so intricately distinguished in relation to one another, are to him polluting. Thus for any ego within the system the threatening non-structure against which barriers must be erected lies below. The sad wit of pollution as it comments on bodily functions symbolises descent in the caste structure by contact with faeces, blood and corpses.

The Coorgs shared with other castes this fear of what is outside and below. But living in their mountain fastness they were also an isolated community, having only occasional and controllable contact with the world around. For them the model of the exits and entrances of the human body is a doubly apt symbolic focus of fears for their minority standing in the larger society. Here I am suggesting that when rituals express anxiety about the body's orifices the sociological counterpart of this anxiety is a care to protect the political and cultural unity of a minority group. The Israelites were always in their history a hard-pressed minority. In their beliefs all the bodily issues were polluting, blood, pus, excreta, semen, etc. The threatened boundaries of their body politic would be well mirrored in their care for the integrity, unity and purity of the physical body.

The Hindoo caste system, while embracing all minorities, embraces them each as a distinctive, cultural sub-unit. In any given locality, any sub-caste is likely to be a minority. The purer and higher its caste status, the more of a minority it must be. Therefore the revulsion from touching corpses and excreta does not merely express the order of caste in the system as a whole. The anxiety about bodily margins expresses danger to group survival.

That the sociological approach to caste pollution is much more convincing than a psychoanalytic approach is clear when we consider what the Indian's private attitudes to defecation are. In the ritual we know that to touch excrement is to be defiled and that the latrine cleaners stand in the lowest grade of the caste hierarchy. If this pollution rule expressed individual anxieties we would expect Hindoos to be controlled and secretive about the act of defecation. It comes as a considerable shock to read that slack disregard is their normal attitude, to such an extent that pavements, verandahs and public places are littered with faeces until the sweeper comes along.

'Indians defecate everywhere. They defecate, mostly beside the railway tracks. But they also defecate on the beaches; they defecate on the streets; they never look for cover. . . . These squatting figures—to the visitor, after a time, as eternal and emblematic as Rodin's Thinker—are never spoken of; they are never written about; they are not mentioned in novels or stories; they do not appear in feature films or documentaries. This might be regarded as part of a permissible prettifying intention. But the truth is that Indians do not see these squatters and might even, with complete sincerity, deny that they exist.'

(Naipaul, chapter 3)

Rather than oral or anal eroticism it is more convincing to argue that caste pollution represents only what it claims to be. It is a symbolic system, based on the image of the body, whose primary concern is the ordering of a social hierarchy.

It is worth using the Indian example to ask why saliva and genital excretions are more pollution-worthy than tears. If I can fervently drink his tears, wrote Jean Genêt, why not the so limpid drop on the end of his nose? To this we can reply: first that nasal secretions are not so limpid as tears. They are more like treacle than water. When a thick rheum oozes from the eye it is no more apt for poetry than nasal rheum. But admittedly clear, fast-running tears are the stuff of romantic poetry: they do not defile. This is partly because tears are naturally pre-empted by the symbolism of washing. Tears are like rivers of moving water. They purify, cleanse, bathe the eyes, so how can they pollute? But more significantly tears are not related to the bodily functions of digestion or procreation. Therefore their scope for symbolising social relations and social processes is narrower. This is evident when we reflect on caste structure. Since place in the hierarchy of purity is biologically transmitted, sexual behaviour is important for preserving the purity of caste. For this reason, in higher castes, boundary pollution focusses particularly on sexuality. The caste membership of an individual is determined by his mother, for though she may have married into a higher caste, her children take their caste from her. Therefore women are the gates of entry to the caste. Female purity is carefully guarded and a woman who is known to have had sexual intercourse with a man of lower caste is brutally punished. Male sexual purity does not carry this responsibility. Hence male promiscuity is a lighter matter. A mere ritual bath is enough to cleanse a man from sexual contact with a low caste woman. But his sexuality does not entirely escape the burden of worry which boundary pollution attaches to the body. According to Hindoo belief a sacred quality inheres in semen, which should not be wasted. In a penetrating essay on female purity in India (1963) Yalman says:

'While caste purity must be protected in women and men may be allowed much greater freedom, it is, of course, better for the men not to waste the sacred quality contained in their semen. It is well-known that they are exhorted not merely to avoid low-caste women, but all women (Carstairs 1956, 1957;

Gough 1956). For the loss of semen is the loss of this potent stuff . . . it is best never to sleep with women at all.'

Both male and female physiology lend themselves to the analogy with the vessel which must not pour away or dilute its vital fluids. Females are correctly seen as, literally, the entry by which the pure content may be adulterated. Males are treated as pores through which the precious stuff may ooze out and be lost, the whole system being thereby enfeebled.

A double moral standard is often applied to sexual offences. In a patrilineal system of descent wives are the door of entry to the group. In this they hold a place analogous to that of sisters in the Hindoo caste. Through the adultery of a wife impure blood is introduced to the lineage. So the symbolism of the imperfect vessel appropriately weighs more heavily on the women than on the men.

If we treat ritual protection of bodily orifices as a symbol of social preoccupations about exits and entrances, the purity of cooked food becomes important. I quote a passage on the capacity of cooked food to be polluted and to carry pollution (in an unsigned review article on Pure and Impure, *Contributions to Indian Sociology*, III, July 1959, p. 37)

'When a man uses an object it becomes part of him, participates in him. Then, no doubt, this appropriation is much closer in the case of food, and the point is that appropriation precedes absorption, as it accompanies the cooking. Cooking may be taken to imply a complete appropriation of the food by the household. It is almost as if, before being "internally absorbed" by the individual, food was, by cooking, collectively predigested. One cannot share the food prepared by people without sharing in their nature. This is one aspect of the situation. Another is that cooked food is extremely permeable to pollution.'

This reads like a correct transliteration of Indian pollution symbolism regarding cooked food. But what is gained by proffering a descriptive account as if it were explanatory? In India the cooking process is seen as the beginning of ingestion, and therefore cooking is susceptible to pollution, in the same way as eating. But why is this complex found in India and in parts of Polynesia and in Judaism and other places, but not wherever humans sit down to eat? I suggest that food is not likely to be polluting at all unless the external boundaries of the social system are under pressure. We can go further to explain why the actual cooking of the food in India must be ritually pure. The purity of the castes is correlated with an elaborate hereditary division of labour between castes. The work performed by each caste carries a symbolic load: it says something about the relatively pure status of the caste in question. Some kinds of labour correspond with the excretory functions of the body, for example that of washermen, barbers, sweepers, as we have seen. Some professions are involved with bloodshed or alcoholic liquor, such as tanners, warriors, toddy tappers. So they are low in the scale of purity in so far as their occupations are at variance with Brahminic ideals. But the point at which food is prepared for the table is the point at which the interrelation of the purity structure and the occupational structure needs to be set straight. For food is produced by the combined efforts of several castes of varying degrees of purity: the blacksmith, carpenter, ropemaker, the peasant. Before being admitted to the body some clear symbolic break is needed to express food's separation from necessary but impure contacts. The cooking process, entrusted to pure hands, provides this ritual break. Some such break we would expect to find whenever the production of food is in the hands of the relatively impure.

These are the general lines on which primitive rituals must be related to the social order and the culture in which they are found. The examples I have given are crude, intended to exemplify a broad objection to a certain current treatment of ritual themes. I add one more, even cruder, to underline my point. Much literature has been expended by psychologists on Yurok pollution ideas (Erikson, Posinsky). These North Californian Indians who lived by fishing for salmon in the Klamath river, would seem to have been obsessed by the behaviour of liquids, if their pollution rules can be said to express an obsession. They are careful not to mix good water with bad, not to urinate into rivers, not to mix sea and fresh water, and so on. I insist that these rules cannot imply obsessional neuroses, and they cannot be interpreted unless the fluid formlessness of their highly competitive social life be taken into account (Dubois).

To sum up. There is unquestionably a relation between individual preoccupations and primitive ritual. But the relation is not the simple one which some psychoanalysts have assumed. Primitive ritual draws upon individual experience, of course.

This is a truism. But it draws upon it so selectively that it cannot be said to be primarily inspired by the need to solve individual problems common to the human race, still less explained by clinical research. Primitives are not trying to cure or prevent personal neuroses by their public rituals. Psychologists can tell us whether the public expression of individual anxieties is likely to solve personal problems or not. Certainly we must suppose that some interaction of the kind is probable. But that is not at issue. The analysis of ritual symbolism cannot begin until we recognise ritual as an attempt to create and maintain a particular culture, a particular set of assumptions by which experience is controlled.

Any culture is a series of related structures which comprise social forms, values, cosmology, the whole of knowledge and through which all experience is mediated. Certain cultural themes are expressed by rites of bodily manipulation. In this very general sense primitive culture can be said to be autoplastic. But the objective of these rituals is not negative withdrawal from reality. The assertions they make are not usefully to be compared to the withdrawal of the infant into thumb-sucking and masturbation. The rituals enact the form of social relations and in giving these relations visible expression they enable people to know their own society. The rituals work upon the body politic through the symbolic medium of the physical body.

8

Internal Lines

IN THE BEGINNING of this century it was held that primitive ideas
of contagion had nothing to do with ethics. This was how a
special category of ritual called magic came to be instituted for
scholarly discussion. If pollution ritual could be shown to have
some connection with morals its place would have been squarely
within the field of religion. To complete our survey of how
early religion has fared at the hands of early anthropology, it
remains to show that pollution has indeed much to do with
morals.

It is true that pollution rules do not correspond closely to
moral rules. Some kinds of behaviour may be judged wrong
and yet not provoke pollution beliefs, while others not thought
very reprehensible are held to be polluting and dangerous. Here
and there we find that what is wrong is also polluting. Pollution
rules only high-light a small aspect of morally disapproved behaviour. But we still need to ask whether pollution touches on
morals in an arbitrary fashion or not.

To answer this we need to consider moral situations more
closely and to think of the relation between conscience and social
structure. By and large the private conscience and the public
code of morals influence one another continually. As David
Pole says:

'The public code that makes and moulds the private conscience
is remade and moulded by it in turn. . . . In the real recipro-
city of the process, public code and private conscience flow
together: each springs from and contributes to the other,
channels it and is channelled. Both alike are redirected and
enlarged.' (pp. 91-2)

It is not usually necessary to make much distinction between the two. However, we find that we cannot understand this field of pollution unless we enter the sphere which lies between that behaviour which an individual approves for himself and what he approves for others; between what he approves as a matter of principle and what he vehemently desires for himself here and now in contradiction of the principle; between what he approves in the long term and what he approves in the short term. In all this there is scope for discrepancy.

We should start by recognising that moral situations are not easy to define. They are more usually obscure and contradictory than clear cut. It is the nature of a moral rule to be general, and its application to a particular context must be uncertain. For instance, the Nuer believe that homicide within the local community and incest are wrong. But a man may be led into breaking the homicide rule by following another rule of approved behaviour. Since Nuer are taught from boyhood to defend their rights by force, any man may unintentionally kill a fellow villager in a fight. Again the rules of prohibited sexual relationship are complicated and genealogical reckoning in some directions is rather sketchy. A man may easily not be sure whether a particular woman stands to him in a prohibited degree of relationship or not. So there can often be more than one view of what action is right, because of disagreement about what is relevant to the moral judgment and about the estimated consequence of an act. Pollution rules, by contrast with moral rules are unequivocal. They do not depend on intention or a nice balancing of rights and duties. The only material question is whether a forbidden contact has taken place or not. If pollution dangers were placed strategically along the crucial points in the moral code, they could theoretically reinforce it. However, such a strategic distribution of pollution rules is impossible, since the moral code by its nature can never be reduced to something simple, hard and fast.

However, as we look more closely at the relation between pollution and moral attitudes we shall discern something very like attempts to buttress a simplified moral code in this way. To stay with the same tribe, Nuer cannot always tell whether they have committed incest or not. But they believe that incest brings misfortune in the form of skin disease, which can be averted by sacrifice. If they know they have incurred the risk they can have the sacrifice performed; if they reckon the degree of relationship was very distant, and the risk therefore slight, they can leave the matter to be settled *post hoc* by the appearance or non-appearance of the skin disease. Thus pollution rules can serve to settle uncertain moral issues.

Nuer attitudes to the contacts which they consider dangerous are not necessarily disapproving. They would be horrified at a case of incest between mother and son, but many of the relationships which are prohibited to them arouse no such condemnation. A 'little incest' is something which could happen between the best families at any time. Similarly, they regard the effects of adultery to be dangerous to the injured husband; he is liable to contract pains in his back when he subsequently has intercourse with his wife, and this can only be averted by a sacrifice for which the beast should be provided by the adulterer. But although an adulterer may be killed without compensation if caught red-handed, the Nuer do not seem to disapprove of adultery in itself. One gets the impression that pursuing other men's wives is seen as a risky sport in which any man may normally be tempted to indulge (Evans-Pritchard, 1951).

Now it is the same Nuer who have the pollution fears and make the moral judgments: the anthropologist does not believe that the often lethal punishments for incest and adultery are externally imposed on them by their severe god in the interests of maintaining the social structure. The integrity of the social structure is very much at issue when breaches of the adultery and incest rules are made, for the local structure consists entirely of categories of persons defined by incest regulations, marriage payments and marital status. To have produced such a society the Nuer have evidently needed to make complicated rules about incest and adultery, and to maintain it they have underpinned the rules by threats of the danger of forbidden contacts. These rules and sanctions express the public conscience, the Nuer when they are thinking in general terms. Any particular case of adultery or incest interests the Nuer in another way. Men seem to identify with adulterers more than with aggrieved husbands. Their feelings of moral disapproval are not very much engaged on behalf of matrimony and the social structure when confronted by a particular case. Hence one cause of the discrepancy between pollution rules and moral judgments. It suggests that pollution rules can have another socially useful function—that of marshalling moral disapproval when it lags. The Nuer husband, disabled or even dying of adultery pollution, is recognised as the victim of the adulterer: unless the latter pays his fine and provides the sacrifice he will have a death on his hands.

Another general point is suggested by this example. We have given instances of behaviour which the Nuer often regard as morally neutral, and yet which they believe sets off dangerous manifestations of power. There are also types of behaviour which Nuer regard as thoroughly reprehensible, and which are not thought to incur automatic danger. For example, it is a positive duty for a son to honour his father, and acts of filial disrespect are thought to be very wrong. But unlike lack of respect towards parents-in-law, they are not visited with automatic punishment. The social difference between the two situations is that a man's own father as head of the joint family and controller of its herds is in a strong economic position for asserting his superior status, while the father-in-law or mother-in-law is not. This accords with the general principle that when the sense of outrage is adequately equipped with practical sanctions in the social order, pollution is not likely to arise. Where, humanly speaking, the outrage is likely to go unpunished, pollution beliefs tend to be called in to supplement the lack of other sanctions.

To sum up, if we could extract from the whole volume of Nuer behaviour certain kinds of behaviour which are condemned by them as wrong, we would have a map of their moral code. If we could make another map of their pollution beliefs, we would find that it touches the outline of morality here and there, but is by no means congruent with it. A large part of their pollution rules concern etiquette between husband and wife and between in-laws. The punishments which are thought to fall on those who break these rules can be accounted for by Radcliffe-Brown's formula of social value: the rules express the value of marriage in that society. They are specific pollution rules, such as one forbidding a wife to drink the milk of the cows which have been paid for her marriage. Such rules do not coincide with moral rules, though they may well express approval of general attitudes (such as respect towards one's husband's herds). These rules only relate indirectly to the moral code insofar as they draw attention to the value of behaviour which has some bearing on the structure of society, the code of morality being itself related to that same social structure.

Then there are other pollution rules which touch the moral code more closely, such as those forbidding incest or homicide within the local community. The fact that pollution beliefs provide a kind of impersonal punishment for wrongdoing affords a means of supporting the accepted system of morality. The Nuer examples suggest the following ways in which pollution beliefs can uphold the moral code:

(i) When a situation is morally ill-defined, a pollution belief can provide a rule for determining *post hoc* whether infraction has taken place, or not.

(ii) When moral principles come into conflict, a pollution rule can reduce confusion by giving a simple focus for concern.

(iii) When action that is held to be morally wrong does not provoke moral indignation, belief in the harmful consequences of a pollution can have the effect of aggravating the seriousness of the offence, and so of marshalling public opinion on the side of the right.

(iv) When moral indignation is not reinforced by practical sanctions, pollution beliefs can provide a deterrent to wrongdoers.

This last point can be expanded. In a small scale society the machinery of retribution is never likely to be very strong or very certain in its action. We find that pollution beliefs reinforce it in two distinct ways. Either the transgressor is himself held to be the victim of his own act, or some innocent victim takes the brunt of the danger. This we would expect to vary in a regular manner. In any social system there may be some strongly held moral norms whose breach cannot be punished. For example, when self-help is the only way of righting wrong, people are banded for protection into groups which pursue vengeance for their members. In such a system there can easily be no way of exacting vengeance when a murder has been committed within the group itself; deliberately to kill or even to outlaw a fellow-member would be to offend against the strongest principle of all. In such cases we commonly find that pollution danger is expected to fall on the head of the fratricide.

This is a very different problem from the pollution whose dangers fall, not on the head of the transgressor but on the innocent. We saw that the innocent Nuer husband is the one whose life is risked when his wife commits adultery. There are many variations on this theme. Often it is the guilty wife, sometimes it is the injured husband, often it is his children whose lives are endangered. The adulterer himself is not often thought to risk danger, though the Ontong Javanese hold this belief (Hogbin, p. 153). In the case of the fratricide above, moral indignation is not lacking. The problem is a practical question of how to punish rather than one of how to arouse moral fervour against the crime. The danger replaces active human punishment. In the case of adultery pollution the belief that the innocent are in danger helps to brand the delinquent and to rouse moral fervour against him. So in this case pollution ideas strengthen the demand for active human punishment.

It is outside the scope of this study to collect and compare a large number of examples. But here is a field which it would be interesting to tackle by documentary research. What are the precise circumstances in which adultery pollution is thought to endanger the injured husband, the unborn or the living children, or the delinquent or innocent wife? Whenever danger follows secret adultery in a social system in which someone has the right to claim damages if adultery is known, the pollution belief acts as a *post hoc* detector of the crime. This fits the Nuer case above. Another example comes in a text given by a Nyakyusa husband:

> 'If I have always been all right and strong and I find that I get tired walking and hoeing, I think: "What is it? See, always I was all right and now I am very tired." My friends say: "It is a woman, you have lain with one who was menstruating." And if I eat food and start diarrhoea, they say: "It is women, they have committed adultery!" My wives deny it. We go to divine and one is caught; if she agrees, that's that, but if she denied it, formerly we went to the poison ordeal. The woman drank alone not I. If she vomited then I was defeated, the woman was good, but if it caught her then her father paid me one cow.' (Wilson, p. 133)

Similarly, when it is believed that a woman will miscarry if she has committed adultery while pregnant, and that her infant will die if she commits adultery while suckling it, someone may have a case for blood-compensation for every confessed adultery. If girls are normally married before puberty and are expected to go from pregnancy to delivery and from delivery to a three or four year suckling period, and then to new pregnancy again, the husband is theoretically insured against infidelity until her menopause. Furthermore, the behaviour of the wife herself is, in this way, very heavily sanctioned by risks for her children and for her own life in the hour of labour. All this makes good sense. Pollution beliefs here uphold marital relations. But we are still no nearer answering why it should be in some cases the husband who is the victim and in others the wife in child labour or the children, and in others again, as among the Bemba, the innocent party, whether the husband or the wife, who is automatically endangered.

The answer must lie in a minute examination of the distribution of rights and duties in marriage and the various interests and advantages of each party. The varying incidence of danger allows moral judgment to point to different individuals: if the wife herself is endangered, even to the point of risking her life in child labour, indignation is summoned against her seducer. This suggests a society in which the wife is less likely to get a beating for her misconduct. If the husband's life is endangered then blame presumably falls on the wife or her lover. As a long shot (more for the sake of making some suggestion that can be tested than with much confidence in its soundness), could it be that the danger falls on the wife when, for some reason or other she cannot be openly chastised? Perhaps because the presence of her kinsfolk in the village protects her? Then we might expect that in the opposite case, when the danger falls on the husband, this gives him an added excuse for giving her a good beating, or at least summons the opinion of the community against her loose behaviour. Here I suggest that a society where marriage is stable and where wives are kept in control may be one in which the danger of adultery may fall on the aggrieved husband.

So far we have discussed four ways in which pollution tends to support moral values. The fact that pollutions are easier to cancel than moral defects gives us another set of situations. Some pollutions are too grave for the offender to be allowed to survive. But most pollutions have a very simple remedy for undoing their effects. There are rites of reversing, untying, burying, washing, erasing, fumigating, and so on, which at a small cost of time and effort can satisfactorily expunge them. The cancelling of a moral offence depends on the state of mind of the offended party and on the sweetness of nursing revenge. The social consequences of some offences ripple out in all directions and can never be reversed. Rites of reconciliation which enact the burial of the wrong have the creative effect of all ritual. They can help to erase memory of the wrong and encourage the growth of right feeling. There must be an advantage for society at large in attempting to reduce moral offences to pollution offences which can be instantly scrubbed out by ritual. Levy-Bruhl, who gave many examples of rituals of purification (1936, Chapter VIII), had the insight to note that the act of restitution itself takes on the status of a rite of annulment. He points out that the law of talion is misunderstood if it is seen merely as meeting a brutal need for vengeance:

> 'To the necessity of a counter-action equal to and like the action, is associated the law of talion . . . because he has suffered an attack, received a wound, undergone a wrong, he feels exposed to an evil influence. A threat of misfortune hangs over him. To reassure himself, to regain calm and security, the evil influence thus released must be stopped, neutralised. Now this result will not be obtained unless the act from which he suffers is annulled by a similar act in the opposite direction. This is precisely what retaliation procures for the primitive.'
>
> (pp. 392-5)

Levy-Bruhl did not make the mistake of supposing that a purely external act is sufficient. He noticed, as anthropologists have continued to notice ever since, the strenuous efforts that are made to bring the inward heart and mind into line with the public act. The contradiction between external behaviour and secret emotions is a frequent source of anxiety and of expected misfortune. This is a new contradiction which can arise from the act of purification itself. We should therefore recognise it as an autonomous pollution in its own right. Levy-Bruhl gives many examples of what he calls the bewitching effects of ill-will (p. 186).

These pollutions, which lurk between the visible act and the invisible thought, are like witchcraft. They are dangers from the crevices of the structure, and like witchcraft their inherent power to harm does not depend either on external action or on any deliberate intention. They are dangerous in themselves.

There are two distinct ways of cancelling a pollution: one is the ritual which makes no enquiry into the cause of the pollution, and does not seek to place responsibility; the other is the confessional rite. On the face of it one would expect these to apply in very different situations. Nuer sacrifice is an example of the first. Nuer associate misfortunes with offences which have brought them about, but they do not seek to relate a particular misfortune to a particular offence. The question is regarded as academic, since the only resource open to them in all cases is the same, sacrifice. An exception is the case of adultery we have mentioned. It is necessary to know the adulterer so that he can produce the beast for sacrifice, and also be mulcted of a fine. Reflecting on this instance, we can suppose that confession, since it always makes precise the nature of the offence and enables blame to be allocated, is a good basis for demanding compensation.

A new kind of relation between pollution and morals emerges when purification alone is taken to be an adequate treatment for moral wrongs. Then the whole complex of ideas including pollution and purification become a kind of safety net which allows people to perform what, in terms of social structure, could be like acrobatic feats on the high wire. The equilibrist dares the impossible and lightly defies the laws of gravity. Easy purification enables people to defy with impunity the hard realities of their social system. For example, the Bemba have such good confidence in their technique of purification from adultery that though they believe adultery has lethal dangers they give reign to their short term desires. This case I discuss in more detail in the next chapter. What is relevant here is the apparently contradictory fear of sex and pleasure in sex which Dr. Richards remarked (pp. 154-5), and the role of purification rites in overcoming the fears. She insists that no Bemba supposes that fear of adultery pollution deters anyone from adultery.

From this we are led to the last point relating pollution to morals. Any complex of symbols can take on a cultural life of its own and even acquire initiative in the development of social institutions. For example, among the Bemba their sex pollution rules would seem on the face of it to express approval of fidelity between husband and wife. In practice divorce is now common and one gets the impression (Richards, 1940) that they turn to divorce and remarriage as a means of avoiding the pollution of adultery. This radical deflection from once held objectives is only possible when other forces of disintegration are at work. We cannot suppose that pollution fears suddenly take the bit between their teeth and gallop off with the social system. But they can ironically provide independent grounds for breaking the moral code which at one time they worked to support.

Pollution ideas can distract from the social and moral aspects of a situation by focussing on a simple material matter. The Bemba believe that pollution of adultery is conveyed through fire. Therefore the careful housewife seems to be obsessed by the problem of protecting her hearth from adulterous and menstrual defilement and from homicides.

'It is difficult to exaggerate the strength of these beliefs or the extent to which they affect daily life. In a village at cooking time young children are sent here and there to fetch "new fire" from neighbours who are ritually pure.' (p. 33)

Why their anxieties about sex should have been transferred from bed to board belongs to the next chapter. But why fire needs to be protected depends on the configuration of powers which control their universe. Death, blood and coldness are confronted by their opposites, life, sex and fire. All six powers are dangerous. The three positive powers are dangerous unless separated from one another and are in danger from contact with death, blood or coldness. The act of sex is always to be separated from the rest of life by a rite of purification which only husband and wife can perform for each other. The adulterer is a public danger because his contact defiles all cooking hearths and he cannot be purified. From this we see that anxieties about their social life are only part of the explanation of Bemba sex pollution. To explain why fire (rather than, for instance, salt as among some of their neighbours) should convey pollution we would need to approach the systematic interrelation of the symbols themselves in more detail than is at present possible.

This cursory sketch is as far as I can go on the relation between pollution and morals. To have shown that the relation is far from straightforward is necessary before returning to the idea of society as a complex set of Chinese boxes, each sub-system having little sub-systems of its own, and so on indefinitely for as far as we care to apply the analysis. It is my belief that people really do think of their own social environment as consisting of other people joined or separated by lines which must be respected. Some of the lines are protected by firm physical sanctions. There are churches in which tramps do not sleep on the benches because the vestry-man will call the police. Ultimately India's lower castes used to keep in their place because of similarly effective social sanctions, and all the way up the edifice of caste political and economic forces help to maintain the system. But wherever the lines are precarious we find pollution ideas come to their support. Physical crossing of the social barrier is treated as a dangerous pollution, with any of the consequences we have just examined. The polluter becomes a doubly wicked object of reprobation, first because he crossed the line and second because he endangered others.

9

The System at War with Itself

WHEN THE COMMUNITY is attacked from outside at least the external danger fosters solidarity within. When it is attacked from within by wanton individuals, they can be punished and the structure publicly reaffirmed. But it is possible for the structure to be self-defeating. This has long been a familiar theme for anthropologists (see Gluckman 1963). Perhaps all social systems are built on contradiction, in some sense at war with themselves. But in some cases the various ends which individuals are encouraged to pursue are more harmoniously related than in others.

Sexual collaboration is by nature fertile, constructive, the common basis of social life. But sometimes we find that instead of dependence and harmony, sexual institutions express rigid separation and violent antagonism. So far we have noted a kind of sex pollution which expresses a desire to keep the body (physical and social) intact. Its rules are phrased to control entrances and exits. Another kind of sex pollution arises from the desire to keep straight the internal lines of the social system. In the last chapter we noted how rules control individual contacts which destroy these lines, adulteries, incests and so forth. But these by no means exhaust the types of sexual pollution. A third type may arise from the conflict in the aims which can be proposed in the same culture.

In primitive cultures, almost by definition, the distinction of the sexes is the primary social distinction. This means that some important institutions always rest on the difference of sex. If the social structure were weakly organised, then men and women might still hope to follow their own fancies in choosing and discarding sexual partners, with no grievous consequences for society at large. But if the primitive social structure is strictly articulated, it is almost bound to impinge heavily on the relation between men and women. Then we find pollution ideas enlisted to bind men and women to their allotted roles, as we have shown in the last chapter.

There is one exception we should note at once. Sex is likely to be pollution-free in a society where sexual roles are enforced directly. In such a case anyone who threatened to deviate would be promptly punished with physical force. This supposes an administrative efficiency and consensus which are rare anywhere and specially in primitive societies. As an example we can consider the Walbiri of Central Australia, a people who unhesitatingly apply force to ensure that the sexual behaviour of individuals shall not undermine that part of the social structure which rests upon marital relations (Meggitt). As in the rest of Australia, a great part of the social system depends upon rules governing marriage. The Walbiri live in a hard desert environment. They are aware of the difficulty of community survival and their culture accepts as one of its objectives that all members of the community shall work and be cared for according to their ability and needs. This means that responsibility for the infirm and old falls upon the hale. A strict discipline is asserted throughout the community, young are subject to their seniors, and above all, women are subject to men. A married woman usually lives at a distance from her father and brothers. This means that though she has a theoretical claim to their protection, in practice it is null. She is in the control of her husband. As a general rule if the female sex were completely subject to the male, no problem would be posed by the principle of male dominance. It could be enforced ruthlessly and directly wherever it applied. This seems to be what happens among the Walbiri. For the least complaint or neglect of duty Walbiri women are beaten or speared. No blood compensation can be claimed for a wife killed by her husband, and no one has the right to intervene between husband and wife. Public opinion never reproaches the man who has violently, or even lethally, asserted his authority over his wife. Thus it is impossible for a woman to play off one man against another. However energetically they may try to seduce one another's wives the men are in perfect accord on one point. They are agreed that they should never allow their sexual desires to give an individual woman bargaining power or scope for intrigue.

These people have no beliefs concerning sex pollution. Even

menstrual blood is not avoided, and there are no beliefs that contact with it brings danger. Although the definition of married status is important in their society it is protected by overt means. Here there is nothing precarious or contradictory about male dominance.

No constraints are imposed on individual Walbiri men. They seduce one another's women if they get a chance, without showing any special concern for the social structure based on marriage. The latter is preserved by the thorough-going subordination of women to men and by the recognised system of self-help. When a man poaches on another's sexual preserve he knows what he risks, a fight and possible death. The system is perfectly simple. There are conflicts between men, but not between principles. No moral judgment is evoked in one situation which is likely to be contradicted in another. People are held to these particular roles by the threat of physical violence. The previous chapter has suggested that when this threat is uninhibited we can expect the social system to persist without the support of pollution beliefs.

It is important to recognise that male dominance does not always flourish with such ruthless simplicity. In the last chapter we saw that when moral rules are obscure or contradictory there is a tendency for pollution beliefs to simplify or clarify the point at issue. The Walbiri case suggests a correlation. When male dominance is accepted as a central principle of social organisation and applied without inhibition and with full rights of physical coercion, beliefs in sex pollution are not likely to be highly developed. On the other hand, when the principle of male dominance is applied to the ordering of social life but is contradicted by other principles such as that of female independence, or the inherent right of women as the weaker sex to be more protected from violence than men, then sex pollution is likely to flourish. Before we take up this case there is another kind of exception to consider.

We find many societies in which individuals are not coerced or otherwise held strictly to their allotted sexual roles and yet the social structure is based upon the association of the sexes. In these cases a subtle, legalistic development of special institutions provides relief. Individuals can to some extent follow their personal whims, because the social structure is cushioned by fictions of one kind or another.

The political organisation of the Nuer is totally unformulated. They have no explicit institutions of government or administration. Such fluid and intangible political structure as they exhibit is a spontaneous, shifting expression of their conflicting loyalties. The only principle of any firmness which gives form to their tribal life is the principle of genealogy. By thinking of their territorial units as if they represented segments of a single genealogical structure, they impose some order on their political groupings. The Nuer afford a natural illustration of how people can create and maintain a social structure in the realm of ideas and not primarily, or at all, in the external, physical realm of ceremonial, palaces or courts of justice (Evans-Pritchard, 1940).

The genealogical principle which they apply to the political relations of a whole tribe is important to them in another context, at the intimate personal level of claims to cattle and wives. Thus, not only his place in the larger political scheme, but his personal inheritance is determined for a Nuer man by the allegiances defined through marriage. On rights of paternity their lineage structure and their whole political structure depend. Yet the Nuer do not take adultery and desertion so tragically as some other peoples with agnatic lineage systems in which paternity is established by marriage. True a Nuer husband can spear his wife's seducer if he catches him red-handed. But otherwise, if he learns of her infidelity, he can only demand two head of cattle, one for compensation and one for sacrifice—hardly a severe penalty compared with other peoples of whom we read that they used to banish adulterers (Meek, p. 218-9) or enslave them. Or compared with a Bedouin who would not be allowed to raise his head in society again until a dishonoured kinswomen had been killed (Salim, p. 61). The difference is that Nuer legal marriage is relatively invulnerable to the whims of individual partners. Husbands and wives can be allowed to separate and live apart without altering the legal status of their marriage, or of the wife's children (Evans-Pritchard, Chapter III, 1951). Nuer women enjoy a strikingly free and independent status. If one is widowed her husband's brothers have the right to take her in leviratic marriage, to raise seed to the dead man's name. But if she does not choose to accept this arrangement, they cannot force her. She is left free to choose her own lovers. The one security that is guaranteed to the dead man's lineage is that any off-

spring, begotten by whomsoever they may have been, count as
affiliated to that lineage from which the original marriage cattle
were paid. The rule that whoever paid cattle has the right to
the children is the rule which distinguishes legal marriage, practically indefeasible, from conjugal relations. The social structure
rests on the series of legal marriages, established by the transfer
of cattle. Thus it is protected by practical institutional means
from any uncertainty which may threaten from the free behaviour of men and women. By contrast with the stark, unstated
simplicity of their political organisation, Nuer display astonish-
ing legal subtlety in the definition of marriage, concubinage,
divorce and conjugal separation.

It is this development, I suggest, which enables them to organise their social institutions without burdensome beliefs in
sex pollution. It is true that they protect their cattle from
menstruating women, but a man does not have to purify himself
if he touches one. He should merely avoid sexual intercourse
with his wife during her menstrual periods, a rule of respect
which is said to express concern for his unborn children. This
is a very much milder regulation than some rules of avoidance
we shall mention later.

We have earlier noted another example of a legal fiction which
lifts the weight of the social structure from sexual relations. This
is Nur Yalman's discussion of female purity in South India and
Ceylon (1963). Here the purity of women is protected as the gate
of entry to the castes. The mother is the decisive parent for
establishing caste membership. Through women the blood and
purity of the caste is perpetuated. Therefore their sexual purity
is all-important, and every possible whisper of threat to it is
anticipated and barred against. This should lead us to expect an
intolerable life of restriction for women. Indeed this is what
we find for the highest and purest caste of all.

The Nambudiri Brahmins of Malabar are a small, rich, exclusive caste of priestly land-owners. They have remained so by
observing a rule forbidding the division of their estates. In each
family only the eldest son marries. The others can keep lower
caste concubines, but never enter into marriage. As for their
unfortunate womenfolk, strict seclusion is their lot. Few of them
ever marry at all until on their deathbed a rite of marriage
affirms their freedom from the control of their guardians. If they
go out of their houses, their bodies are completely enveloped

in clothing and umbrellas hide their faces. When one of their brothers is married they can watch the celebration through chinks in the walls. Even at her own wedding a Nambudiri woman has to be replaced in the usual public appearance of the bride by a Nayar girl. Only a very rich group could afford to commit its women to a life sentence of barrenness for most and of seclusion for all. This kind of ruthlessness corresponds in its own way to the ruthlessness with which Walbiri men apply their principles.

But though similar ideas about purity of women prevail in the other castes, this hard solution has not been adopted. Orthodox Brahmins, who do not try to keep their patrimonial estates intact and allow their sons to marry, preserve the purity of their women by requiring girls to be married before puberty to suitable husbands. They put strong moral and religious pressures on one another to ensure that every Brahmin girl is properly married before her first menstruation. In other castes if they do not arrange a real marriage before puberty, then a substitute rite of marriage is absolutely required. In middle India she can first be married to an arrow or a wooden pounder. This counts as a first marriage and gives the girls married status so that any misdemeanours of hers can be dealt with in the caste or local courts on the model of a married woman.

Southern Nayar girls are renowned in India for the sexual licence they enjoy. No permanent husband is recognised; the women live at home and have loose relationships with a wide number of men. The caste position of these women and of their children is made ritually secure by a pre-puberty rite of substitute marriage. The man who takes the part of the ritual bridegroom is himself of appropriate caste status and he provides ritual paternity for the girl's future offspring. Should a Nayar girl at any time be thought to have had contact with a man of lower caste, she would be as brutally punished as a woman of the Nambudiri Brahmins. But, apart from guarding against such a lapse, her life is as free and uncontrolled as perhaps any woman's within the caste system, and a great contrast with her Nambudiri 'neighbours' secluded regime. The fiction of first marriage has lifted from her much of the burden of protecting the purity of the blood of the caste.

So much for the exceptions.

We should now look at some examples of social structures

which rest on grave paradox or contradiction. In these cases where no softening legal fictions intervene to protect the freedom of the sexes exaggerated avoidances develop around sexual relations.

In different cultures the accepted theories of cosmic power give more or less explicit place to sexual energy. In the cultures of Hindu India, for example, and of New Guinea, the symbolism of sex occupies a central place in the cosmology. But among African Nilotes, by contrast, the sexual analogy seems to be much less developed. It would be vain to pretend to relate the broad lines of these metaphysical variations to differences in the social organisation. But within any such a cultural region we find interesting minor variations on the theme of sexual symbolism and pollution. These we can and should try to correlate with other local variations.

New Guinea is an area where fear of sexual pollution is a cultural characteristic (Read, 1954). But within the same cultural idiom a great contrast separates the way the Arapesh of Sepik River and the Mae Enga of the Central Highlands handle the theme of sexual difference. The former, it seems, try to create a total symmetry between the sexes. All power is thought of on the model of sexual energy. Femininity is only dangerous to men as masculinity is to women. Females are life-giving and in pregnancy they nourish their children with their own blood; once the children are born males nourish them with life-giving blood drawn for the purpose from the penis. Margaret Mead emphasises that equal watchfulness is necessary from both sexes on their own dangerous powers. Each sex approaches the other with deliberate control (1940).

The Mae Enga, on the other hand, do not look for any symmetry. They fear female pollution for their males and for all male enterprises, and there is no question of a balance between two kinds of sexual danger and powers (Meggitt, 1964). For such differences we can tentatively look for sociological correlations.

The Mae Enga live in an area of dense population. Their local organisation is based on the clan, a compact, well-defined military and political unit. The men of the clan choose their wives from other clans. Thus they marry foreigners. The rule of clan exogamy is common enough. Whether it imports strain and difficulty into the marriage situation depends on how exclusive, localised and rivalrous are the intermarrying clans. In the Enga case they are not only foreigners but traditional enemies. The Mae Enga men are individually involved in an intense competition for prestige. They fiercely compete to exchange pigs and valuables. Their wives are chosen from the very outsiders with whom they habitually exchange pigs and shells and with whom they habitually fight. So for each man his male affines are also likely to be his ceremonial exchange partners (a competitive relationship) and their clan is the military enemy of his own clan. Thus the marital relation has to bear the tensions of the strongly competitive social system. The Enga belief about sex pollution suggests that sexual relations take on the character of a conflict between enemies in which the man sees himself as endangered by his sexual partner, the intrusive member of the enemy clan. There is a strongly held belief that contacts with women weaken male strength. So preoccupied are they with avoiding female contact that the fear of sexual contamination effectively reduces the amount of commerce between the sexes. Meggitt has evidence that adultery used to be unknown, and divorces practically unheard of.

From early boyhood the Enga are taught to shun female company, and they go into periodic seclusion to purify themselves from female contact. The two dominant beliefs in their culture are the superiority of the male principle and its vulnerability to female influence. Only a married man can risk sexual intercourse because the special remedies for protecting virility are only available to married men. But even in marriage men fear sexual activity and would seem to reduce it to the minimum necessary for procreation. Above all they fear menstrual blood:

'They believe that contact with it or with a menstruating woman will, in the absence of appropriate counter-magic, sicken a man and cause persistent vomiting, "kill" his blood so that it turns black, corrupt his vital juices so that his skin darkens and hangs in folds as his flesh wastes, permanently dull his wits, and eventually lead to a slow decline and death.'

Dr. Meggitt's own view is that 'The Mae equation of femininity, sexuality and peril' is to be explained by the attempt to found marriage on an alliance which spans the most competitive relations in their highly competitive social system.

'Until recently clans fought constantly over scarce land resources, pig-thefts and failure to meet debts, and in any given clan most of the men lost in battle have been killed by its immediate neighbours. At the same time, because of the rugged mountainous terrain, propinquity has been a significant variable in determining actual marriage choices. Thus there is a relatively high correlation between interclan marriage and homicide frequencies with regard to the nearness of clans. The Mae recognise this concomitance in a crude way when they say: "We marry the people we fight".' (Meggitt, 1963)

We noted that the Mae Enga fear of female pollution contrasts with the belief in the balanced power and danger from both sexes that appears in the culture of the Mountain Arapesh. It is very interesting to note further that the Arapesh disapprove of local exogamy. If a man should marry a woman of the plains Arapesh he observes elaborate precautions to cool·off her more dangerous sexuality.

'If he marries one, he should not marry her hastily but permit her to remain about the house for several months growing accustomed to him, cooling down the possible passion of slight acquaintance and strangeness. Then he may copulate with her, and watch. Do his yams prosper? Does he find game when he goes hunting? If so, all is well. If not, let him abstain from relationship with this dangerous, oversexed woman still many more moons, lest the part of his potency, his own physical strength, the ability to feed others, which he most cherishes, should be permanently injured.' (Mead, 1940, p. 345)

This example would seem to support Meggitt's view that local exogamy in the strained and competitive conditions of Enga life imports a heavy load of strain into their marriage. If this is so then the Enga could presumably be free of their very inconvenient beliefs if they could relieve their anxieties at source. But this is an utterly impractical suggestion. It would mean either giving up their violently competitive exchanges with rival clans, or their exogamous marriages—either stop fighting or stop marrying the sisters of the men they fight. Either choice would mean a major readjustment to their social system. In practice and in historical fact, when such an adjustment came from outside, with the coming of missionary preaching on sex and of the Australian administration's pax on fighting, the Enga gave up their beliefs in the danger of female sex quite easily.

The contradiction which the Enga strive to overcome by

avoidance is the attempt to build marriage on enmity. But another difficulty perhaps more common in primitive societies arises from a contradiction in the phrasing of male and female roles. If the principle of male domination is elaborated absolutely consistently, it need not necessarily contradict any other basic principles. We have mentioned two very different instances in which male dominance is applied with ruthless simplicity. But the principle runs into trouble if there is any other principle which protects women from physical control. For this gives women scope to play off one man against another, and so to confound the principle of male dominance.

The whole society is especially likely to be founded on a contradiction if the system is one in which men define their status in terms of rights over women. If there is free competition between the men, this gives scope for a discontented woman to turn to her husband's or her guardian's rivals, gain new protectors and new allegiance, and so to dissolve into nothing the structure of rights and duties which had formerly been built around her. This sort of contradiction in the social system arises only if there is no *de facto* possibility of coercing women. For example, it does not appear if there is a centralised political system which throws the weight of its authority against women. Where the legal system can be exerted against women, they cannot make havoc of the system. But a centralised political system is not one in which male status is mainly phrased in terms of rights over women.

The Lele are an example of a social system which is continually liable to founder on the contradiction that female manoeuvring gives to male dominance. All male rivalries are expressed in the competition for wives. A man with no wife is below the bottom rung of the status ladder. With one wife he can get a start, by begetting and thus qualifying for entry to remunerative cult associations. With a daughter born to him he can start claiming the services of a son-in-law. With several daughters, as many betrothed sons-in-law and above all with granddaughters, he is high up on the ladder of privilege and esteem. This is because women whom he has engendered are women he can offer in marriage to other men. And so he builds up a following of men. Every mature man could hope to acquire two or three wives, and in the meanwhile young men had to wait in bachelorhood. Polygyny in itself made the competition for wives intense. But the various other ways in which male success in the men's sphere was hitched to the control of women would be complicated to relate here (see Douglas, 1963). Their whole social life was dominated by an institution for paying compensation by transferring rights over women. The net effect was that women were treated, from one aspect, as a kind of currency in which men claimed and settled debts against one another. Men's mutual indebtedness piled up so that they had staked out claims to unborn girls for generations ahead. A man with no rights over women to transfer was in as parlous a case as a modern business man with no bank account. From a man's point of view women were the most desirable objects their culture had to offer. Since all insults and obligations could be settled by the transfer of rights over women, it was perfectly true to say, as they did, that the only reason they ever went to war was about women.

A little Lele girl would grow up a coquette. From infancy she was the centre of affectionate, teasing, flirting attention. Her affianced husband never gained more than a very limited control over her. He had the right to chastise, certainly, but if he exerted it too violently, and above all if he lost her affection, she could find some pretext for persuading her brothers that her husband neglected her. Infant mortality was high and the miscarriage or death of a child brought the wife's kinsmen sternly to the husband's door asking for an explanation. Since men competed with one another for women there was scope for women to manoeuvre and intrigue. Hopeful seducers were never lacking and no woman doubted that she could get another husband if it suited her. The husband whose wives were faithful until middle age had to be very attentive, both to the wife and to her mother. Quite an elaborate etiquette governed marital relations, with many occasions on which big or small gifts were due from the husband. When the wife was pregnant or sick or newly delivered, he had to be assiduous in arranging proper medical care. A woman who was known to be dissatisfied with her life would soon be courted—and there were various ways open to her by which she could take initiative for ending her marriage.

I have said enough to show why Lele men should be anxious about their relations with women. Although in some contexts they thought of women as desirable treasures, they spoke of them also as worthless, worse than dogs, unmannerly, ignorant, fickle, unreliable. Socially, women were indeed all these things. They were not in the least interested in the men's world in which they and their daughters were swapped as pawns in men's games of prestige. They were cunning in exploiting the opportunities that came their way. If they connived, mother and daughter together could wreck any plans that they disliked. So ultimately men had to assert their vaunted dominance by charming, coaxing and flattering. There was a special wheedling tone of voice they used for women.

The Lele attitude to sex was compounded of enjoyment, desire for fertility and recognition of danger. They had every reason for desiring fertility, as I have shown, and their religious cults were directed towards this end. Sexual activity was held to be in itself dangerous, not for the partners to it, but dangerous for the weak and the sick. Anyone coming fresh from sexual intercourse should avoid the sick, lest by the indirect contact their fever should increase. New-born babies would be killed by such a contact. Consequently yellow raffia fronds were hung at the entrance of a compound to warn all responsible persons that a sick person or new born baby was within. This was a general danger. But there were special dangers for men. A wife had the duty of cleaning her husband after sexual intercourse and then of washing herself before she touched the cooking. Each married woman kept a little pot of water hidden in the grass outside the village where she could wash in secret. It should be well hidden and out of the way, for if a man were to trip on that pot by chance, his sexual vigour would be weakened. If she neglected her ablution and he were to eat food cooked by her, he would lose his virility. These are just the dangers following legitimate sexual intercourse. But a menstruating woman could not cook for her husband or poke the fire, lest he fall ill. She could prepare the food, but when it came to approaching the fire she had to call a friend in to help. These dangers were only risked by men, not by other women or children. Finally, a menstruating woman was a danger to the whole community if she entered the forest. Not only was her menstruation certain to wreck any enterprise in the forest that she might undertake, but it was thought to produce unfavourable conditions for men. Hunting would be difficult for a long time after, and rituals based on forest plants would have no efficacy. Women found these rules extremely irksome, specially as they were regularly short-handed and late in their planting, weeding, harvesting and fishing.

The danger of sex was also controlled by rules which protected male enterprises from female pollution and female enterprises from male pollution. All ritual had to be protected from female pollution, the male officiants (women were generally excluded from cult affairs) abstaining from sexual intercourse the night before. The same for warfare, hunting, tapping palms for wine. Similarly women should abstain from sexual intercourse before planting ground nuts or maize, fishing, making salt or pottery. These fears were symmetrical for men and for women. The generally stipulated condition for handling any great ritual crisis was to call for sexual abstinence from the whole village. Thus when twins were born, or when a twin from another village entered for the first time, or in the course of important anti-sorcery or fertility rituals, the villagers would hear it announced night after night 'Each man his mat alone, each woman her mat alone'. At the same time they would hear it announced 'Let no one quarrel tonight. Or if you must quarrel, don't quarrel in secret. Let us hear the noise, so that we can impose a fine'. Quarrelling was, like sexual intercourse, regarded as being destructive of the proper ritual condition of the village. It spoilt ritual and hunting. But quarrelling was always bad, while sexual intercourse was only bad on certain (rather frequent) occasions.

The Lele anxiety about the ritual dangers of sex I attribute to the real disruptive role allotted to sex in their social system. Their men created a status ladder whose successive stages they mounted as they acquired control over more and more women. But they threw the whole system open to competition and so allowed women a double role, as passive pawns, and again as active intriguers. Individual men were right to fear that individual women would spoil their plans, and fears of the dangers of sex only too accurately reflect its working in their social structure.

Female pollution in a society of this type is largely related to the attempt to treat women simultaneously as persons and as the currency of male transactions. Males and females are set off as belonging to distinct, mutually hostile spheres. Sexual antagonism inevitably results and this is reflected in the idea that each sex constitutes a danger to the other. The particular dangers which female contact threatens to males express the contradiction of trying to use women as currency without reducing them to slavery. If ever it was felt in a commercial culture that money is the root of all evil, the feeling that women are the root of all evils to Lele men is more justified. Indeed the story of the Garden of Eden touched a deep chord of sympathy in Lele male breasts. Once told by the missionaries, it was told and retold round pagan hearths with smug relish.

The Yurok of Northern California have more than once interested anthropologists and psychologists by the radical nature of their ideas of purity and impurity as we have said. Theirs is a dying culture. When Professor Robins studied the Yurok language in 1951 there were only about six Yurok-speaking adults left alive. This seems to have been another example of a highly competitive, acquisitive culture. Men's minds were preoccupied with acquiring wealth in the form of prestige-carrying shell-money, rare feathers and pelts and imported obsidian blades. Apart from those who had access to the routes along which the foreign valuables were traded, the normal way of acquiring wealth was by being quick to avenge wrongs and by demanding compensation. Every insult had its price, more or less standardised. There was latitude for haggling since the price was finally agreed *ad hoc*, according to the value a man set on himself and the backing he could muster among his close kinsmen (Kroeber). Adulteries of wives and marriages of daughters were important sources of wealth. A man who pursued other men's wives would pour out his fortune in adultery compensation.

The Yurok so much believed that contact with women would destroy their powers of acquiring wealth that they held that women and money should never be brought into contact. Above all it was felt to be fatal to future prosperity for a man to have sexual intercourse in the house where he kept his strings of shell money. In the winter, when it was too cold to be out of doors, they seem to have abstained altogether. For Yurok babies tended to be born at the same time of year—nine months after the first warm weather. Such a rigorous separation of business and pleasure tempted Walter Goldschmidt to compare Yurok values with those of the Protestant ethic. The exercise involved him in a highly specious stretching of the notion of capitalist economy, so that it would embrace both the salmon-fishing Yurok and 16th century Europe. He showed that a high value on chastity, parsimony and pursuit of wealth characterised both societies. He also laid great emphasis on the fact that the Yurok could be classed as primitive capitalists since they admitted private control of the means of production, unlike most other primitive peoples. Well, it is true that Yurok individuals laid claim to fishing and berry sites and that these could in the last resort be transferred from one individual to another in settlement of debts. But this is a very special pleading if it is to be the basis for classing the economy as capitalist. Such transfers only took place exceptionally as a kind of foreclosure when a man had no shell money or other movable wealth to pay big debts, and it is obvious that there was no regular market in real estate. The debts which Yurok normally incurred were not commercial debts but debts of honour. Cora Dubois has given an illuminating account of neighbouring peoples where the fierce competition for prestige was played in a sphere more or less insulated from the subsistence sphere of the economy. It is much more significant for understanding their idea of female pollution that for the Yurok men there was a real sense in which pursuit of wealth and of women were contradictory.

We have traced this Delilah complex, the belief that women weaken or betray, in various extreme forms among the New Guinea Mae Enga and among the Lele of the Congo and the Yurok Indians of California. Where it occurs we find that men's anxieties about women's behaviour are justified and that the situation of male/female relations is so biased that women are cast as betrayers from the start.

It is not always the men who are afraid of sex pollution. For the sake of symmetry we should look at one example where it is the women who behave as if sexual activity were highly dangerous. Audrey Richards says that the Bemba of Northern Rhodesia behave as if they were obsessed by fear of sexual impurity. But she notes that this is culturally standardised behaviour, and in fact no fears seem to check their individual freedom. At the cultural level the fear of sexual intercourse seems dominant to an extent 'which cannot be exaggerated'. At the personal level there is 'the open pleasure in sex relations which the Bemba express' (1956, p. 154).

In other places sexual pollution is incurred by direct contact, but here it is held to be mediated by contact with fire. There is no danger in seeing or touching a sexually active, unpurified person, someone hot with sex, as the Bemba say. But let such a person come near a fire and any food cooked on those flames is dangerously contaminated.

It takes two to have sexual intercourse, but only one to cook a meal. By supposing the pollution to be transmitted through cooked food responsibility is firmly fixed on the Bemba women. A Bemba woman has to be continually alert to protect her cooking hearth from the contact of any adult who may have had sexual intercourse without ritual purification. The danger is lethal. Any child who eats food cooked on a contaminated fire may die. A Bemba mother is kept busy putting out suspect fires and lighting new, pure ones.

Although the Bemba believe that all sexual activity is dangerous, the bias of their beliefs points to adultery as the real, practical danger. A married couple are able to administer ritual purification, each for the other, after every sexual act. But an adulterer cannot be purified unless he can ask his wife to help, as it is not a solo rite.

Dr. Richards does not tell us how the impurity of adultery is ever cancelled or how, in the long term, the adulteress feeds her own children. These beliefs, she assures us, do not deter them from adultery. So dangerous adulterers are thought to be at large. Though they may try conscientiously not to touch hearths where infant food is being cooked, they always remain a potential public danger.

Notice that in this society the women are more anxious than the men about sexual pollution. If their children die (and the infant mortality rate is very high) they can be blamed for carelessness by the men. In Nyasaland among the Yao and Cewa a similar complex of beliefs is expressed concerning pollution of salt. All three tribes reckon descent in the female line, and in all three tribes the men are supposed to leave their natal village and join each the village of his wife. This gives a pattern of village structure by which a core of lineally related females attracts men from other villages to settle as their husbands. The future of the village as a political unit depends on keeping these male outsiders living there. But we would expect the men to have much less interest in building a stable marriage. The same rule of matrilineal succession turns their interest to their sisters' children. Though the village is built on the conjugal tie, the matrilineal lineage is not. The men are brought to the village by marriage, the women are born in it.

Throughout Central Africa the idea of the good village which grows and endures is a value strongly held by men and women. But the women have a double interest in keeping their husbands. A Bemba woman fulfils her most satisfying role when, in middle life, as a matriarch in her own village she can expect to grow old surrounded by her daughters and her daughters' daughters. But if a Bemba man finds the early years of marriage irksome, he will simply leave his wife and go home (Richards, p. 41). Moreover, if all the men go, or even half of them, the village is no longer viable as an economic unit. The division of labour puts Bemba women in a particularly dependent position. Indeed, in a region where it is now common for fifty per cent of the adult males to be absent on labour migration, Bemba villages suffered more disintegration than villages of other tribes in Northern Rhodesia (Watson).

The teaching of Bemba girls in their puberty ceremonies helps us to relate these aspects of social structure and of women's ambitions to their fears of sex pollution. Dr. Richards records that the girls are strictly indoctrinated with the need to behave submissively to their husbands; interesting since they are reputed to be particularly overbearing and difficult to manage. The candidates are humiliated while their husbands' virility is extolled. This makes good sense if we consider the Bemba husband's role as analogous, in a contrary way, to that of the Mae Enga wife. He is alone and an outsider in his wife's village. But he is a man and not a woman. If he is not happy he goes away and there is an end to it. He cannot be chastised like a runaway wife. There are no legal adjustments by which the fiction of a legal marriage can be preserved without insisting on the reality. His physical presence in his wife's village is more important to that village than the rights he gains in marriage are to himself, and he cannot be browbeaten into staying there. If the Enga wife is a Delilah, he is Samson in the camp of the Philistines. If he feels humiliated he can bring the pillars of society tumbling down, for if all the husbands were to rise up and go the village would be ruined. No wonder that the women are anxious to flatter and cajole him. No wonder they would like to protect against the effects of adultery. The husband appears not to be dangerous or sinister, but shy, liable to be frightened off, needing to be convinced of his own masculinity and of the dangers thereof. He needs to be assured that his wife is looking after him, standing by to purify him, watching over the fire. He can do nothing without her, not even approach his own ancestral spirits. In her self-imposed anxieties about sex pollution the Bemba wife appears as the opposite number of the Mae Enga husband. Both find in the marriage situation anxieties concerning the structure of the wider society. If the Bemba woman did not want to stay at home and become an influential matron there, if she were prepared to follow her husband meekly to his village, she could relieve her anxiety about sexual pollution.

In all the examples quoted of this kind of pollution, the basic problem is a case of wanting to have your cake and eat it. The Enga want to fight their enemy clans but yet to marry with their clanswomen. The Lele want to use women as the pawns of men, and yet will take sides with individual women against other men. The Bemba women want to be free and independent and to behave in ways which threaten to wreck their marriages, and yet they want their husbands to stay with them. In each case the dangerous situation which has to be handled with washings and avoidances has in common with the others that the norms of behaviour are contradictory. The left hand is fighting the right hand, as in the Trickster myth of the Winnebago.

Is there any reason why all these examples of the social system at war with itself are drawn from sexual relations? There are many other contexts in which we are led into contradictory behaviour by the normal canons of our culture. National income policy is one modern field in which this sort of analysis could easily be applied. Yet pollution fears do not seem to cluster round contradictions which do not involve sex. The answer may be that no other social pressures are potentially so explosive as those which constrain sexual relations. We can come to sympathise with St. Paul's extraordinary demand that in the new Christian society there should be neither male nor female.

The cases we have considered may throw some light on the exaggerated importance attached to virginity in the early centuries of Christianity. The primitive church of the Acts in its treatment of women was setting a standard of freedom and equality which was against the traditional Jewish custom. The barrier of sex in the Middle East at that time was a barrier of oppression, as St. Paul's words imply:

'Baptised into Christ, you have put on Christ: there can be

neither Jew, nor Greek, nor bond nor free, there can be neither
male nor female, for you are all one man in Christ Jesus.'

(Gal. 3. 28)

In its effort to create a new society which would be free, un-
bounded and without coercion or contradiction, it was no doubt
necessary to establish a new set of positive values. The idea that
virginity had a special positive value was bound to fall on good
soil in a small persecuted minority group. For we have seen that
these social conditions lend themselves to beliefs which symbolise the body as an imperfect container which will only be
perfect if it can be made impermeable. Further, the idea of the
high value of virginity would be well-chosen for the project of
changing the role of the sexes in marriage and in society at
large (Wangermann). The idea of woman as the Old Eve, together with fears of sex pollution, belongs with a certain specific
type of social organisation. If this social order has to be changed,
the Second Eve, a virgin source of redemption crushing evil
underfoot is a potent new symbol to present.

The System Shattered and Renewed

NOW TO CONFRONT our opening question. Can there be any people who confound sacredness with uncleanness? We have seen how the idea of contagion is at work in religion and society. We have seen that powers are attributed to any structure of ideas, and that rules of avoidance make a visible public recognition of its boundaries. But this is not to say that the sacred is unclean. Each culture must have its own notions of dirt and defilement which are contrasted with its notions of the positive structure which must not be negated. To talk about a confused blending of the Sacred and the Unclean is outright nonsense. But it still remains true that religions often sacralise the very unclean things which have been rejected with abhorrence. We must, therefore, ask how dirt, which is normally destructive, sometimes becomes creative.

First, we note that not all unclean things are used constructively in ritual. It does not suffice for something to be unclean for it to be treated as potent for good. In Israel it was unthinkable that unclean things, such as corpses and excreta could be incorporated into the Temple ritual, but only blood, and only blood shed in sacrifice. Among the Oyo Yoruba where the left hand is used for unclean work and it is deeply insulting to proffer the left hand, normal rituals sacralise the precedence of the right side, especially dancing to the right. But in the ritual of the great Ogboni cult initiates must knot their garments on the left side and dance only to the left (Morton-Williams, p. 369). Incest is a pollution among the Bushong, but an act of ritual incest is part of the sacralisation of their king and he claims that he is the filth of the nation: '*Moi, ordure, nyec*' (Vansina, p. 103). And so on. Though it is only specific individuals on specified occasions who can break the rules, it is still important to ask why these dangerous contacts are often required in rituals.

One answer lies in the nature of dirt itself. The other lies in the nature of metaphysical problems and of particular kinds of reflections which call for expression.

To deal with dirt first. In the course of any imposing of order, whether in the mind or in the external world, the attitude to rejected bits and pieces goes through two stages. First they are recognisably out of place, a threat to good order, and so are regarded as objectionable and vigorously brushed away. At this stage they have some identity: they can be seen to be unwanted bits of whatever it was they came from, hair or food or wrappings. This is the stage at which they are dangerous; their half-identity still clings to them and the clarity of the scene in which they obtrude is impaired by their presence. But a long process of pulverizing, dissolving and rotting awaits any physical things that have been recognised as dirt. In the end, all identity is gone. The origin of the various bits and pieces is lost and they have entered into the mass of common rubbish. It is unpleasant to poke about in the refuse to try to recover anything, for this revives identity. So long as identity is absent, rubbish is not dangerous. It does not even create ambiguous perceptions since it clearly belongs in a defined place, a rubbish heap of one kind or another. Even the bones of buried kings rouse little awe and the thought that the air is full of the dust of corpses of bygone races has no power to move. Where there is no differentiation there is no defilement.

> 'They outnumber the living, but where are all their bones?
> For every man alive there are a million dead,
> Has their dust gone into earth that it is never seen?
> There should be no air to breathe, with it so thick,
> No space for wind to blow or rain to fall:
> Earth should be a cloud of dust, a soil of bones,
> With no room even for our skeletons.
> It is wasted time to think of it, to count its grains,
> When all are alike and there is no difference in them.'
>
> (S. Sitwell, Agamemnon's Tomb)

In this final stage of total disintegration, dirt is utterly undifferentiated. Thus a cycle has been completed. Dirt was created by the differentiating activity of mind, it was a by-product of the creation of order. So it started from a state of non-differentiation; all through the process of differentiating its role was to threaten the distinctions made; finally it returns to its true indiscriminable character. Formlessness is therefore an apt symbol of beginning and of growth as it is of decay.

On this argument everything that is said to explain the revivifying role of water in religious symbolism can also apply to dirt:

> 'In water everything is "dissolved", every "form" is broken up, everything that has happened ceases to exist; nothing that was before remains after immersion in water, not an outline, not a "sign", not an event. Immersion is the equivalent, at the human level, of death at the cosmic level, of the cataclysm (the Flood) which periodically dissolves the world into the primeval ocean. Breaking up all forms, doing away with the past, water possesses this power of purifying, of regenerating, of giving new birth. . . . Water purifies and regenerates because it nullifies the past, and restores—even if only for a moment—the integrity of the dawn of things.' (Eliade, 1958, p. 194)

In the same book Eliade goes on to assimilate with water two other symbols of renewal which we can, without labouring the point, equally associate with dust and corruption. One is symbolism of darkness and the other orgiastic celebration of the New Year (pp. 398-9).

In its last phase then, dirt shows itself as an apt symbol of creative formlessness. But it is from its first phase that it derives its force. The danger which is risked by boundary transgression is power. Those vulnerable margins and those attacking forces which threaten to destroy good order represent the powers inhering in the cosmos. Ritual which can harness these for good is harnessing power indeed.

So much for the aptness of the symbol itself. Now for the living situations to which it applies, and which are irremediably subject to paradox. The quest for purity is pursued by rejection. It follows that when purity is not a symbol but something lived, it must be poor and barren. It is part of our condition that the purity for which we strive and sacrifice so much turns out to be hard and dead as a stone when we get it. It is all very well for the poet to praise winter as the

'Paragon of art,
That kills all forms of life and feeling
Save what is pure and will survive.'
(Roy Campbell)

It is another thing to try and make over our existence into an unchanging lapidary form. Purity is the enemy of change, of ambiguity and compromise. Most of us indeed would feel safer if our experience could be hard-set and fixed in form. As Sartre wrote so bitterly of the anti-semite:

'How can anyone choose to reason falsely? It is simply the old yearning for impermeability . . . there are people who are attracted by the permanence of stone. They would like to be solid and impenetrable, they do not want change: for who knows what change might bring? . . . It is as if their own existence were perpetually in suspense. But they want to exist in all ways at once, and all in one instant. They have no wish to acquire ideas, they want them to be innate . . . they want to adopt a mode of life in which reasoning and the quest for truth play only a subordinate part, in which nothing is sought except what has already been found, in which one never becomes anything else but what one already was.' (1948)

This diatribe implies a division between ours and the rigid black and white thinking of the anti-semite. Whereas, of course, the yearning for rigidity is in us all. It is part of our human condition to long for hard lines and clear concepts. When we have them we have to either face the fact that some realities elude them, or else blind ourselves to the inadequacy of the concepts.

The final paradox of the search for purity is that it is an attempt to force experience into logical categories of non-contradiction. But experience is not amenable and those who make the attempt find themselves led into contradiction.

Where sexual purity is concerned it is obvious that if it is to imply no contact between the sexes it is not only a denial of sex, but must be literally barren. It also leads to contradiction. To wish all women to be chaste at all times goes contrary to other wishes and if followed consistently leads to inconveniences of the kind to which Mae Enga men submit. High-born girls of 17th century Spain found themselves in a dilemma in which dishonour stood on either horn. St. Theresa of Avila was brought up in a society in which the seduction of a girl had to be avenged by her brother or father. So if she received a lover she risked dishonour and the lives of men. But her personal honour required her to be generous and not to withhold herself from her lover, as it was unthinkable to shun lovers altogether. There are many other examples of how the quest for purity creates problems and some curious solutions.

One solution is to enjoy purity at second hand. Something of a vicarious satisfaction gave its aura, no doubt, to the respect for virginity in early Christendom, gives extra zest to the Nambudiri Brahmins when they enclose their sisters, and enhances the prestige of Brahmins among lower castes in general. In certain chiefdoms the Pende of the Kasai expect their chiefs to live in sexual continence. Thus one man conserves the wellbeing of the chiefdom on behalf of his polygamous subjects. To ensure no lapse on the part of the chief, who is admittedly past his prime when installed, his subjects fix a penis sheath on him for life (de Sousberghe).

Sometimes the claim to superior purity is based on deceit. The adult men of the Chagga tribe used to pretend that at initiation their anus was blocked for life. Initiated men were supposed never to need to defecate, unlike women and children who remained subject to the exigency of their bodies (Raum). Imagine the complications into which this pretence led Chagga men. The moral of all this is that the facts of existence are a chaotic jumble. If we select from the body's image a few aspects which do not offend, we must be prepared to suffer for the distortion. The body is not a slightly porous jug. To switch the metaphor, a garden is not a tapestry; if all the weeds are removed, the soil is impoverished. Somehow the gardener must preserve fertility by returning what he has taken out. The special kind of treatment which some religions accord to anomalies and abominations to make them powerful for good is like turning weeds and lawn cuttings into compost.

This is the general outline for an answer to why pollutions are often used in renewal rites.

Whenever a strict pattern of purity is imposed on our lives it is either highly uncomfortable or it leads into contradiction if closely followed, or it leads to hypocrisy. That which is negated is not thereby removed. The rest of life, which does not tidily fit the accepted categories, is still there and demands attention. The body, as we have tried to show, provides a basic scheme for all symbolism. There is hardly any pollution which does not have some primary physiological reference. As life is in the body it cannot be rejected outright. And as life must be affirmed, the most complete philosophies, as William James put it, must find some ultimate way of affirming that which has been rejected.

'If we admit that evil is an essential part of our being and the key to the interpretation of our life, we load ourselves down with a difficulty that has always proved burdensome in philosophies of religion. Theism, wherever it has erected itself into a systematic philosophy of the universe, has shown a reluctance to let God be anything less than All-in-All . . . at variance with popular theism (is a philosophy) which is frankly pluralistic . . . the universe compounded of many original principles . . . God is not necessarily responsible for the existence of evil. The gospel of healthy-mindedness casts its vote distinctly for this pluralistic view. Whereas the monistic philosopher finds himself more or less bound to say, as Hegel said, that everything actual is rational, and that evil, as an element dialectically required must be pinned in, and kept and consecrated and have a function awarded to it in the final system of truth, healthy-mindedness refuses to say anything of the sort. Evil, it says, is emphatically irrational, and *not* to be pinned in, or preserved, or consecrated in any final system of truth. It is a pure abomination to the Lord, an alien unreality, a waste element, to be sloughed off and negated . . . the ideal, so far from being coextensive with the actual, is a mere extract from the actual, marked by its deliverance from all contact with this diseased, inferior, excrementitious stuff.

Here we have the interesting notion . . . of there being elements of the universe which may make no rational whole in conjunction with the other elements, and which, from the point of view of any system which those elements make up, can only be considered so much irrelevance and accident—so much "dirt" as it were, and matter out of place.' (p. 129)

This splendid passage invites us to compare dirt-affirming with dirt-rejecting philosophies. If it were possible to make such a comparison between primitive cultures, what would we expect to find? Norman Brown has suggested (see Chapter 8) that primitive magic is an escape from reality, on a par with infantile sexual fantasies. If this were right we should expect to find the majority of primitive cultures lined up with Christian Science, the only example of healthy-mindedness which William James described. But instead of consistent dirt-rejecting, we find the extraordinary examples of dirt-affirmation with which this chapter started. In a given culture it seems that some kinds of behaviour or natural phenomena are recognised as utterly wrong by all the principles which govern the universe. There are different kinds of impossibilities, anomalies, bad mixings and abominations. Most of the items receive varying degrees of condemnation and avoidance. Then suddenly we find that one of the most abominable or impossible is singled out and put into a very special kind of ritual frame that marks it off from other experience. The frame ensures that the categories which the normal avoidances sustain are not threatened or affected in any way. Within the ritual frame the abomination is then handled as a source of tremendous power. On William James's terms, such ritual mixing up and composting of polluting things would provide the basis of 'more complete religion'.

> 'It may indeed be that no religious reconciliation with the absolute totality of things is possible. Some evils, indeed, are ministerial to higher forms of good, but it may be that there are forms of evil so extreme as to enter into no good system whatsoever, and that, in respect of such evil, dumb submission or neglect to notice is the only practical resource. . . . But . . . since the evil facts are as genuine parts of nature as the good ones, the philosophic presumption should be that they have some rational significance, and that systematic healthy-mindedness, failing as it does to accord to sorrow, pain and death any positive and active attention whatever, is formally less complete than systems that try at least to include these elements in their scope. The completest religions would therefore seem to be those in which the pessimistic elements are best developed . . .'
>
> (p. 161)

Here we seem to have the outline of a programme for comparative religion. It would be to their own cost that anthropologists should neglect their duty of drawing up a taxonomy of tribal religions. But we find that it is not a simple matter to work out the best principles for distinguishing the 'incomplete and optimistic' religions from the 'more complete and pessimistic' ones. Problems of method loom large. Obviously one would have to be meticulously scrupulous in cataloguing all the ritual avoidances in any particular religion and in leaving nothing out. Beyond that, what other rules would objective scholarship need, to distinguish different kinds of religion according to these general criteria?

The answer is that the task is utterly beyond the scope of objective scholarship. This is not for the technical reason that the fieldwork is missing. Indeed, the scantier the field research the more practicable the comparative project appears. The reason lies in the nature of the material itself. All live religions are many things. The formal ritual of public occasions teaches one set of doctrine. There is no reason to suppose that its message is necessarily consistent with those taught in private rituals, or that all public rituals are consistent with one another, nor all private rituals. There is no guarantee that the ritual is homogeneous, and if it is not, only the subjective intuition of the observer can say whether the total effect is optimistic or pessimistic. He may follow some rules for arriving at his conclusion; he may decide to add up each side of a balance sheet of evil-rejecting and evil-affirming rites, scoring each one equally. Or he may weight the score according to the importance of the rites. But whatever rule he follows he is bound to be arbitrary. And even then he has only come to the end of the formal ritual. There are other beliefs which may not be ritualised at all, and which may altogether obscure the message of the rites. People do not necessarily listen to their preachers. Their real guiding beliefs may be cheerfully optimistic and dirt-rejecting while they appear to subscribe to a nobly pessimistic religion.

If I were to decide where the Lele culture should be classed on William James' scheme, I would be in a quandary. These are a people who are very pollution-conscious in secular and ritual affairs. Their habitual separating and classifying comes out nowhere so clearly as in their approach to animal food. Most of their cosmology and much of their social order is reflected in their animal categories. Certain animals and parts of animals are appropriate for men to eat, others for women, others for children, others for pregnant women. Others are regarded as totally inedible. One way or another the animals which they reject as unsuitable for human or female consumption turn out to be ambiguous according to their scheme of classification. Their animal taxonomy separates night from day animals; animals of the above (birds, squirrels and monkeys) from animals of

the below: water animals and land animals. Those whose behaviour is ambiguous are treated as anomalies of one kind or
another and are struck off someone's diet sheet. For instance,
flying squirrels are not unambiguously birds nor animals, and
so they are avoided by discriminating adults. Children might
eat them. No woman worthy of the name would eat them, and
men only when driven by hunger. No penalties sanction this
attitude.

One can schematise their main divisions as two concentric
circles. The circle of human society encloses men as hunters
and diviners, women and children and also, anomalously, animals which live in human society. These non-humans in the
village are either domesticated animals, dogs and chickens, or
unwanted parasites, rats and lizards. It is unthinkable to eat
dogs, rats or lizards. Human's meat should be the game brought
in from the wild by the hunters' arrows and traps. Chickens
present something of a problem in casuistry which the Lele
solve by regarding it unseemly for women to eat chicken, though
the meat is possible and even good food for men. Goats, which
are recently introduced, they rear for exchange with other tribes
and do not eat.

All this squeamishness and discrimination would, if consistently carried through, make their culture look like a dirt-rejecting one. But it is what happens in the final count that matters.
For the main part, their formal rituals are based on discrimination of categories, human, animal, male, female, young, old,
etc. But they lead through a series of cults which allow their
initiates to eat what is normally dangerous and forbidden,
carnivorous animals, chest of game and young animals. In an
inner cult a hybrid monster, which in secular life one would
expect them to abhor, is reverently eaten by initiates and taken
to be the most powerful source of fertility. At this point one
sees that this is, after all, to continue the gardening metaphor, a
composting religion. That which is rejected is ploughed back
for a renewal of life.

The two worlds, human and animal, are not at all independent. Most of the animals exist, as the Lele think, to be the
quarry of Lele hunters. Some animals, burrowing or nocturnal,
or water-loving, are spirit animals who have a special connection
with the non-animal inhabitants of the animal world, the spirits.
On these spirits humans depend for prosperity and fertility

and healing. The normal movement is for humans to go out and get what they need from the animal sphere. Animals and spirits characteristically are shy of humans and do not come out spontaneously into the human world. Men, as hunters and diviners, exploit both aspects of this other world, for meat and medicines. Women, as weak and vulnerable, are those who specially need male action in the other world. Women avoid spirit animals and do not eat their meat. Women are never hunters and only become diviners if they are born as, or bear, twins. In the interaction of the two worlds their role is passive, and yet they particularly need the help of the spirits, since women are prone to barrenness, or, if they conceive, to miscarriage, and the spirits can provide remedies.

Apart from this normal relation of male attack and male ritual on behalf of women and children, there are two kinds of mediating bridges between the humans and the wild. One is for evil and the other for good. The dangerous bridge is made by a wicked transfer of allegiance by humans who become sorcerers. They turn their back on their own kind and run with the hunted, fight against the hunters, work against diviners to achieve death instead of healing. They have moved across to the animal sphere and they have caused some animals to move in from the animal to the human sphere. These latter are their carnivorous familiars, who snatch chickens from the human village and do the sorcerers' work there.

The other ambiguous mode of being is concerned with fertility. It is the nature of humans to reproduce with pain and danger and their normal births are single. By contrast, it is thought that animals are naturally fecund; they reproduce without pain or danger and their normal births occur in couples or in larger litters. When a human couple produce twins or triplets they have been able to break through the normal human limitations. In a way they are anomalous, but in the most auspicious possible way. They have a counterpart in the animal world and this is the benign monster to which Lele pay formal cult, the pangolin or scaly ant-eater. Its being contradicts all the most obvious animal categories. It is scaly like a fish, but it climbs trees. It is more like an egg-laying lizard than a mammal, yet it suckles its young. And most significant of all, unlike other small mammals its young are born singly. Instead of running away or attacking, it curls in a modest ball and waits for the hunter to pass. The human twin parents and the forest pangolin, both are ritualised as sources of fertility. Instead of being abhorred and utterly anomalous, the pangolin is eaten in solemn ceremony by its initiates who are thereby enabled to minister fertility to their kind.

This is a mystery of mediation from an animal sphere which parallels the many fascinating human mediators described by Eliade in his account of Shamanism. In their descriptions of the 'pangolin's behaviour and in their attitude to its cult, Lele say things which uncannily recall passages of the Old Testament, interpreted in the Christian tradition. Like Abraham's ram in the thicket and like Christ, the pangolin is spoken of as a voluntary victim. It is not caught, but rather it comes to the village. It is a kingly victim: the village treats its corpse as a living chief and requires the behaviour of respect for a chief on pain of future disaster. If its rituals are faithfully performed the women will conceive and animals will enter hunters' traps and fall to their arrows. The mysteries of the pangolin are sorrowful mysteries:

'Now I will enter the house of affliction,' they sing as initiates carry its corpse round the village. No more of its cult songs were told to me, except this tantalising line. This cult has obviously very many different kinds of significance. Here I limit myself to commenting on two aspects: One is the way it achieves a union of opposites which is a source of power for good; the other is the seemingly voluntary submission of the animal to its own death.

In Chapter 1, I explained why, for the purposes of studying pollution, I would need a broader approach to religion. Defining it as belief in spiritual beings is too narrow. Above all the subject of this chapter is impossible to discuss except in the light of men's common urge to make a unity of all their experience and to overcome distinctions and separations in acts of at-onement. The dramatic combination of opposites is a psychologically satisfying theme full of scope for interpretation at varying levels. But at the same time any ritual which expresses the happy union of opposites is also an apt vehicle for essentially religious themes. The Lele pangolin cult is only one example of which many more could be cited, of cults which invite their initiates to turn round and confront the categories on which their whole surrounding culture has been built up and to recognise them for the fictive, man-made, arbitrary creations that they are. Throughout their daily, and especially their ritual life the Lele are preoccupied with form. Endlessly they enact the discriminations by which their society and its cultural environment exist. and methodically they punish or attribute misfortune to breaches of avoidance rules. The burden of the rules may not be oppressive. But by a conscious effort they respond through them to the idea that creatures of the sky are different in nature from creatures of the earth, so that it is held dangerous for a pregnant woman to eat the latter and nourishing for her to eat the former, and so on. As they prepare to eat they visibly enact the central discriminations of their cosmos no less than the ancient Israelites enacted a liturgy of holiness.

Then comes the inner cult of all their ritual life, in which the initiates of the pangolin, immune to dangers that would kill uninitiated men, approach, hold, kill and eat the animal which in its own existence combines all the elements which Lele culture keeps apart. If they could choose among our philosophies the one most congenial to the moments of that rite, the pangolin initiates would be primitive existentialists. By the mystery of that rite they recognise something of the fortuitous and conventional nature of the categories in whose mould they have their experience. If they consistently shunned ambiguity they would commit themselves to division between ideal and reality. But they confront ambiguity in an extreme and concentrated form. They dare to grasp the pangolin and put it to ritual use, proclaiming that this has more power than any other rites. So the pangolin cult is capable of inspiring a profound meditation on the nature of purity and impurity and on the limitation on human contemplation of existence.

Not only does the pangolin overcome the distinctions in the universe. Its power for good is released by its dying and this it seems to take on itself deliberately. If their religion were all of a piece we might from the foregoing class the Lele as a dirt-affirming religion and expect them to face affliction with resignation, and to make death the occasion of comforting rituals of atonement and renewal. But the metaphysical notions which are all very well in the separate ritual frame of the pangolin cult are another matter when actual death has struck a member of the family. Then the Lele utterly reject the death that has occurred.

It is often said that in this African tribe or that the people do not recognise the possibility of natural death. The Lele are not fools. They recognise that life must come to an end. But if matters were to take their natural course they would expect everyone to live out his natural span and to sink slowly from senility to the grave. When this happens they rejoice, for such an old man or woman has triumphed over all the pitfalls that lay in the way and achieved completion. But this rarely happens. Most people, according to Lele are struck down by sorcery long before they reach their goal. And sorcery does not belong in the natural order of things as Lele see it. Sorcery was a late-coming afterthought, more an accident in creation. In this aspect of their culture they are a good example of the healthy-mindedness which William James described. For the Lele evil is not to be included in the total system of the world, but to be expunged without compromise. All evil is caused by sorcery. They can clearly visualise what reality would be like without sorcery and they continually strive to achieve it by eliminating sorcerers.

A strong millenial tendency is implicit in the way of thinking of any people whose metaphysics push evil out of the world of reality. Among the Lele the millenial tendency bursts into flame in their recurrent anti-sorcery cults. When a new cult arrives it burns up for the time being the whole apparatus of their traditional religion. The elaborate system of anomalies rejected and affirmed which their cults present is regularly superseded by the latest anti-sorcery cult which is nothing less than an attempt to introduce the millenium at once (see Douglas in Middleton & Winter).

Thus we have to reckon with two tendencies in Lele religion: one ready to tear away even the veils imposed by the necessities of thought and to look at reality direct; the other a denial of necessity, a denial of the place of pain and even of death in reality. So William James' problem is turned into the question of which tendency is the stronger.

If the place of the pangolin cult in their world view is what I have described, one would expect it to be slightly orgiastic, a temporary destruction of apollonian form. Perhaps in its origin its feast of communion was a more dionysiac occasion. But there is nothing remotely uncontrolled about Lele rites. They make no use of drugs, dances, hypnosis or any of the arts by which the conscious control of the body is relaxed. Even the one type of diviner who is supposed to be in direct trance communion with the spirits of the forest, and who sings to them all night when they visit him, sings in a staid, austere style. These people are much more concerned with what their religion can deliver in the way of fertility, cures and hunting success than in perfecting man and achieving religious union in the fullest sense. Most of their rites are truly magic rites, performed for the sake of a specific cure or on the eve of a particular hunt, and intended to yield an immediate tangible success. Most of the time the Lele diviners seem no better than a lot of Aladdins rubbing their magic lamps and expecting marvels to take shape. Only their initiation rites into this cult give a glimpse of another level of religious insight. But the teaching of these rites is overlain by the passionate absorption of the people in sorcery and anti-sorcery. Strong political and personal issues hang on the outcome of any sorcery accusation. The rites which detect sorcerers or acquit them, defend against them or restore what they have damaged, these are the rites which steal the public interest. Strong social pressures force people to blame each death on sorcery. Thus it is that whatever their formal religion may say about the nature of the universe and about the place of chaos, suffering and disintegration in reality, the Lele are socially committed to a different view. On this view evil belongs outside the normal scheme of things; it is not part of reality. So the Lele seem to wear the controlled smile of Christian Scientists. If they should be classified not according to their cultic practices, but according to the beliefs which periodically overthrow them, they appear to be frankly healthy-minded, dirt-rejecting, untouched by the lesson of the gentle pangolin.

It would be unfair to take the Lele as an example of a people who try to evade the whole subject of death. I cite their case mainly to show the difficulty of assessing a cultural attitude to such things. I learnt very little about their esoteric doctrines because they were carefully guarded secrets of male cult members. This esotericism in itself is relevant. Lele religious secretiveness is a clear contrast with the much more open rules of admission and publicity of the cultic ritual of the Ndembu, living to the south-east of them. If priests for various social reasons keep their doctrines secret, the anthropologist's misreporting is the least of the evils that can follow. Sorcery fears are less likely to overlay religious teaching, if the religious doctrine is more widely published.

To the Lele, then, it seems that the main reflections to which deaths give rise are thoughts of revenge. Any particular death is treated as unnecessary, due to a wicked crime on the part of a depraved anti-social human being. Just as the focus of all pollution symbolism is the body, the final problem to which the perspective of pollution leads is bodily disintegration. Death presents a challenge to any metaphysical system, but the challenge need not be squarely met. I am suggesting that in treating each death as the outcome of an individual act of treachery and human malice the Lele are evading its metaphysical implications. Their pangolin cult suggests a meditation on the inadequacy of the categories of human thought, but only a few are invited to make it and it is not related explicitly to their experience of death.

It may well seem that I have made too much of the Lele pangolin cult. There are no Lele books of theology or philosophy to state the meaning of the cult. The metaphysical implications have not been expressed to me in so many words by Lele, nor did I even eavesdrop on a conversation between diviners covering this ground. Indeed I have recorded (1957) that I started on the cosmic patterning approach to Lele animal symbolism because I was frustrated in my direct enquiries seeking reasons for their food avoidances. They would never say, 'We avoid anomalous animals because in defying the categories of our universe they arouse deep feelings of disquiet.' But on each avoided animal they would launch into disquisitions on its natural history. The full list of anomalies made clear the simple taxonomic principles being used. But the pangolin was always spoken of as the most incredible monster of all. On first hearing it sounded such a fantastic beast that I could not believe in its existence. On asking why it should be the focus of a fertility cult, I was again frustrated: this was a mystery of the ancestors, way back long ago.

What kind of evidence for the meaning of this cult, or of any cult, can be sensibly demanded? It can have many different levels and kinds of meaning. But the one on which I ground my argument is the meaning which emerges out of a pattern in which the parts can incontestably be shown to be regularly related. No one member of the society is necessarily aware of the whole pattern, any more than speakers are able to be explicit about the linguistic patterns they employ. Luc de Heusch has analysed my material and shown that the pangolin concentrates in its being more of the discriminations central to Lele culture than I myself had realised. I can perhaps justify my interpretation of why they ritually kill and eat it by showing that in other primitive religions similar metaphysical perspectives are recorded. Furthermore, systems of belief are not likely to survive unless they offer reflections on a more profound plane than used to be credited to primitive cultures.

Most religions promise by their rites to make some changes in external events. Whatever promises they make, death must somehow be recognised as inevitable. It is usual to expect that the greatest metaphysical development goes with the most pessimism and contempt of the good things of this life. If religions such as Buddhism teach that individual life is a little thing and that its pleasures are transient and unsatisfying, then they are in a strong philosophical position for contemplating death in the context of the cosmic purpose of an all-pervading Existence. By and large primitive religions and the ordinary layman's acceptance of more elaborate religious philosophies coincide: they are less concerned with philosophy and more interested in the material benefits which ritual and moral conformity can bring. But it follows that those religions which have most emphasised the instrumental effects of their ritual are most vulnerable to disbelief. If the faithful have come to think of rites as means to health and prosperity, like so many magic lamps to be worked by rubbing, there comes a day when the whole ritual apparatus must seem an empty mockery. Somewhere the beliefs must be safeguarded against disappointment or they may not hold assent.

One way of protecting ritual from scepticism is to suppose that an enemy, within or without the community, is continually undoing its good effect. On these lines responsibility may be given to amoral demons or to witches and sorcerers. But this is only a feeble protection for it affirms that the faithful are right in treating ritual as an instrument of their desires, but confesses the weakness of the ritual for achieving its purpose. Thus religions which explain evil by reference to demonology or sorcery are failing to offer a way of comprehending the whole of existence. They come close to an optimistic, healthy-minded,

pluralistic view of the universe. And curiously enough, the prototype of healthy-minded philosophies as William James described
them, Christian Science, was prone to supplement its inadequate
approach to evil by a kind of demonology invented *ad hoc*. I am
grateful to Rosemary Harris for giving me the reference to Mary
Baker Eddy's belief in 'malicious animal magnetism' which she
held accountable for evils she could not ignore (Wilson, 1961,
pp. 126-7).

Another way of protecting the belief that religion can deliver
prosperity here and now is to make ritual efficacy depend on
difficult conditions. On the one hand the rite may be very complicated and difficult to perform: if the least detail gets into the
wrong order, the whole thing is invalid. This is a narrowly instrumental approach, magical in the most pejorative sense. On
the other hand the success of the rite may depend on the moral
conditions being correct: the performer and audience should
be in a proper state of mind, free of guilt, free of ill-will and
so on. A moral requirement for the efficacy of ritual can bind
the believers to the highest purposes of their religion. The prophets of Israel, crying 'Doom, Doom, Doom!' did much more
than provide an explanation of why the rituals failed to give
peace and prosperity. No one who heard them could take a
narrowly magical view of ritual.

The third way is for the religious teaching to change its tack.
In most everyday contexts it tells the faithful that their fields
will prosper and their families flourish if they obey the moral
code and perform the proper ritual services. Then, in another
context, all this pious effort is disparaged, contempt is thrown
on right behaviour, materialistic objectives are suddenly despised.
We cannot say that they suddenly become religions of non-
attachment, promising only disillusionment in this life. But they
travel some way along this path. Thus, for instance, the Ndembu
initiates of Chihamba are made to kill the white spirit that they
have learnt is their grandfather, source of all fertility and health.
Having killed him, they are told they are innocent and must
rejoice (Turner, 1962). Ndembu daily ritual is very intensively
performed as the instrument for gaining good health and good
hunting. Chihamba, their most important cult, is their moment
of disillusion. By it their other cults do not achieve immunity
from discredit. But Turner insists that the object of the
Chihamba rituals is to use paradox and contradiction to express

truths which are inexpressible in any other terms. In Chihamba they confront a more profound reality and measure their objectives by a different standard.

I am tempted to suppose that very many primitive religions which offer material success with one hand, with the other protect themselves from crude experiment by extending their perspective in much the same way. For a narrow focus on material health and happiness makes a religion vulnerable to disbelief. And so we can suppose that the very logic of promises discreditably unfulfilled may lead cult officials to meditate on wider, profounder themes, such as the mystery of evil and of death. If this is true we would expect the most materialistic-seeming cults to stage at some central point in the ritual cycle a cult of the paradox of the ultimate unity of life and death. At such a point pollution of death, treated in a positive creative role, can help to close the metaphysical gap.

We can take for one illustration the death ritual of the Nyakyusa, who live north of Lake Nyasa. They explicitly associate dirt with madness; those who are mad eat filth. There are two kinds of madness, one is sent by God and the other comes from neglect of ritual. Thus they explicitly see ritual as the source of discrimination and of knowledge. Whatever the cause of madness, the symptoms are the same. The madman eats filth and throws off his clothes. Filth is listed as meaning excreta, mud, frogs: 'the eating of filth by madmen is like the filth of death, those faeces are the corpse' (Wilson, 1957, pp. 53, 80-1). So ritual conserves sanity and life: madness brings filth and is a kind of death. Ritual separates death from life: 'the dead, if not separated from the living bring madness on them'. This is a very perspicacious idea of how ritual functions, echoing what we have already seen in Chapter 4, p. 64. Now the Nyakyusa are not tolerant of filth but highly pollution-conscious. They observe elaborate restrictions to avoid contact with bodily rejects which they regard as very dangerous:

'UBANYALI, filth, is held to come from the sex fluids, menstruation and child-birth, as well as from a corpse, and the blood of a slain enemy. All are thought to be both disgusting and dangerous and the sex fluids are particularly dangerous for an infant.' (p. 131)

Contact with menstrual blood is dangerous to a man, specially to a warrior, hence elaborate restrictions on cooking for a man during menstruation.

But in spite of this normal avoidance the central act in the ritual of mourning is actively to welcome filth. They sweep rubbish on to the mourners. 'The rubbish is the rubbish of death, it is dirt. "Let it come now," we say. "Let it not come later, may we never run mad . . ." It means "We have given you everything, we have eaten filth on the hearth." For if one runs mad one eats filth, faeces. . . .' (p. 53). We suspect that there is much more that could be said in the interpretation of this rite. But let us leave it at the point to which the brief remarks of the Nyakusa have taken it : a voluntary embrace of the symbols of death is a kind of prophylactic against the effects of death; the ritual enactment of death is a protection, not against death but against madness (pp. 48-9). On all other occasions they avoid faeces and filth and reckon it a sign of madness not to do so. But in the face of death itself they give up everything, they even claim to have eaten filth as madmen do, in order to keep their reason. Madness will come if they neglect the ritual of freely accepting the corruption of the body; sanity is assured if they perform the ritual.

Another example of death being softened by welcome, if we can put it that way, is the ritual murder by which the Dinka put to death their aged spearmasters. This is the central rite in Dinka religion. All their other rites and bloodily expressive sacrifices pale in significance besides this one which is not a sacrifice. The spearmasters are a hereditary clan of priests. Their divinity, Flesh, is a symbol of life, light and truth. Spearmasters may be possessed by the divinity; the sacrifices they perform and blessings they give are more efficacious than other men's. They mediate between their tribe and divinity. The doctrine underlying the ritual of their death is that the spearmaster's life should not be allowed to escape with his last breath from his dying body. By keeping his life in his body his life is preserved; and the spirit of the spearmaster is thus transmitted to his successor for the good of the community. The community can live on as a rational order because of the unafraid self-sacrifice of its priest.

By reputation among foreign travellers this rite was a brutal suffocation of a helpless old man. An intimate study of Dinka religious ideas reveals the central theme to be the old man's voluntary choosing of the time, manner and place of his death. The old man himself asks for the death to be prepared for him, he asks for it from his people and on their behalf. He is reverently carried to his grave, and lying in it says his last words to his grieving sons before his natural death is anticipated. By his free, deliberate decision he robs death of the uncertainty of its time and place of coming. His own willing death, ritually framed by the grave itself, is a communal victory for all his people (Lienhardt). By confronting death and grasping it firmly he has said something to his people about the nature of life.

The common element in these two examples of death ritual is the exercise of free, rational choice in undergoing death. Something of the same idea is in the self-immolation of the Lele pangolin, and also in the Ndembu ritual killing of Kavula, since this white spirit is not angry but even pleased to be slain. This is yet another theme which death pollution can express if its sign be reversed from bad to good.

Animal and vegetable life cannot help but play their role in the order of the universe. They have little choice but to live as it is their nature to behave. Occasionally the odd species or individual gets out of line and humans react by avoidance of one kind or another. The very reaction to ambiguous behaviour expresses the expectation that all things shall normally conform to the principles which govern the world. But in their own experience as men, people know that their personal conformity is not so certain. Punishments, moral pressures, rules about not touching and not eating, a firm ritual framework, all these can do something to bring man into harmony with the rest of being. But so long as free consent is withheld, so long is the fulfilment imperfect. Here again we can discern primitive existentialists whose escape from the chain of necessity lies only in the exercise of choice. When someone embraces freely the symbols of death, or death itself, then it is consistent with everything that we have seen so far, that a great release of power for good should be expected to follow.

The old spearmaster giving the sign for his own slaying makes a stiffly ritual act. It has none of the exuberance of St. Francis of Assisi rolling naked in the filth and welcoming his Sister Death. But his act touches the same mystery. If anyone held the idea that death and suffering are not an integral part of nature, the delusion is corrected. If there was a temptation to treat ritual as a magic lamp to be rubbed for gaining unlimited riches and power, ritual shows its other side. If the hierarchy of values was crudely material, it is dramatically undermined by paradox and contradiction. In painting such dark themes, pollution symbols are as necessary as the use of black in any depiction whatsoever. Therefore we find corruption enshrined in sacred places and times.

Bibliography

ABERCROMBIE, M. L. JOHNSON, 1960. *The Anatomy of Judgment*. London.

AJOSE, 1957. 'Preventive Medicine and Superstition in Nigeria'. *Africa*, July 1957.

BARTLETT, F. C., 1923. *Psychology and Primitive Culture*. Cambridge. 1932. *Remembering*. Cambridge.

BEATTIE, J., 1960. *Bunyoro, An African Kingdom*. New York. 1964. *Other Cultures*. London.

BERNDT, RONALD, 1951. *Kunapipi, A Study of an Australian Aboriginal Religious Cult*. Melbourne.

BETTELHEIM, B., 1955. *Symbolic Wounds*. Glencoe, Ill.

BLACK, J. S. and CHRYSTAL, G., 1912. *The Life of William Robertson-Smith*. London.

BLACK, M. and ROWLEY, H. H., 1962. (Eds.) *Peake's Commentary on the Bible*. London.

BOHANNAN, P., 1957. *Justice and Judgment among the Tiv*. London.

BROWN, NORMAN, O., 1959. *Life against Death*. London.

BUXTON, JEAN, 1963. Chapter on 'Mandari' in *Witchcraft and Sorcery in East Africa*. (edit. Middleton & Winter). London.

CASSIRER, E., 1944. *An Essay on Man*. Oxford.

CUMMING, E. and J., 1957. *Closed Ranks—an Experiment in Mental Health Education*. Cambridge, Mass.

DE HEUSCH, L., 1964. 'Structure et Praxis Sociales chez les Lele' *L'Homme*, 4. pp. 87-109.

DE SOUSBERGHE, L., 1954. 'Étuis Péniens ou Gaines de Chasteté chez les Ba-Pende.' *Africa*, 24, 3. pp. 214-9.

DOUGLAS, M., 1957. 'Animals in Lele Religious Symbolism'. *Africa*, 27, 1.

1963. *The Lele of the Kasai*. London.

DRIVER, R. S., 1895. *International Critical Commentary on Holy Scriptures of the Old and New Testaments: Deuteronomy*.

Bibliography

DRIVER, R. and WHITE, H. A., 1898. *The Polychrome Bible, Leviticus.* London.

DUBOIS, CORA, 1936. 'The Wealth Concept as an Integrative Factor in Tolowa-Tututni Culture'. Chapter in *Essays in Anthropology, presented to A. L. Kroeber.*

DUMONT, L. and PEACOCK, D., 1959. *Contributions to Indian Sociology,* Vol. III.

DURKHEIM, E., 1912. 1947 edition. Translated by J. Swain, Glencoe, Illinois, *The Elementary Forms of the Religious Life.* Paris. References made to pages in paperback edition, 1961, Collier Books, N.Y.

EHRENZWEIG, A., 1953. *The Psychoanalysis of Artistic Vision and Hearing.* London.

EICHRODT, W., 1933 (first edit.) *Theology of the Old Testament.* Trans. Baker 1961.

ELIADE, M., 1951. *Le Chamanisme* (Trans. 1964). Paris. 1958. *Patterns in Comparative Religion.* London, translated from *Traité d'Histoire des Religions,* 1949.

EPSTEIN, I., 1959. *Judaism.* London.

EVANS-PRITCHARD, E. E., 1934. 'Levy-Bruhl's Theory of Primitive Mentality', *Bulletin of the Faculty of Arts, Cairo, Vol. II, part 1.* 1937. *Witchcraft, Oracles and Magic among the Azande.* Oxford. 1940. *The Nuer.* Oxford. 1951. *Kinship and Marriage among the Nuer.* Oxford. 1956. *Nuer Religion.* Oxford.

FESTINGER, L., 1957. *A Theory of Cognitive Dissonance.* Evanston.

FINLEY, M., 1956. *The World of Odysseus.* Toronto.

FIRTH, R., 1940. 'The Analysis of Mana: an empirical approach'. *Journal of Polynesian Society,* 48. 4. 196. pp. 483-508.

FORTES, M., 1959. *Oedipus and Job in West African Religion.* Cambridge.

FORTES, M. and EVANS-PRITCHARD, G., 1940. *African Political Systems.* Oxford.

FREEDMAN, MAURICE (forthcoming). *Chinese Lineage & Society, Fukien and Kwangtun.*

GELLNER, E., 1962. 'Concepts and Society'. International Sociological Association. *Transactions of the Fifth World Congress of Sociology, Washington, D.C.,* Vol. I.

GENÊT, JEAN, *Journal du Voleur.*

GLUCKMAN, M., 1962. *Essays on the Ritual of Social Relations.* Manchester.

GOFFMAN, E., 1956. *The Presentation of the Self in Everyday Life.* New York.

GOLDSCHMIDT, W., 1951. 'Ethics and the Structure of Society', *American Anthropologist,* 53, 1.

GOODY, J., 'Religion and Ritual: the Definitional Problem'. *British Journal of Sociology.* XII. 2.

GRÖNBECH, V. P. I., 1931. *The Culture of the Teutons,* 2 vols. First printed in Danish, 1909-12.

HARDY, T., 1874. *Far from the Madding Crowd.*

HARPER, ED. B., 1964. *Journal of Asian Studies, XXIII.*

HEGNER, R., ROOT, F., and AUGUSTINE, D., 1929. *Animal Parasitology.* New York and London.

HERZ., J. H., 1935. *The Talmud.*

1938. *Pentateuch & Haftorahs.* London.

HODGEN, MARGARET, 1935. *The Doctrine of Survivals. A chapter in the History of Scientific Method in the Study of Man.* London.

HOGBIN, H. I., 1934. *Law and Order in Polynesia.* London.

HORTON, R., 1961. 'Destiny and the Unconscious in West Africa'. *Africa,* 2, April.

JAMES, WILLIAM, 1901-2. *The Varieties of Religious Experience.* London 1952.

JAMES, E. O., 1938. *Comparative Religion.* Methuen.

KANT, IMMANUEL, 1934. *Immanuel Kant's Critique of Pure Reason,* Norman Kemp Smith, Abridged Edition. Preface to 2nd edit. of Critique of Pure Reason.

KELLOG, S. H., 1841. *The Expositor's Bible.* London.

KOPYTOFF, IGOR, 1964. 'Family and Lineage among the Suku of the Congo', in *The Family Estate in Africa.* Edit. Gray, R. and Gulliver, P.

KRAMER, NOAH, 1956. *From the Tablets of Sumer.* Denver.

KRIGE, E. J. & J. D., 1943. *The Realm of a Rain Queen.* London.

KROEBER, A. L., 1925, *Handbook of the Indians of California.*

LAGRANGE, M. J., 1905. *Études sur les Religions Semitiques.* (2nd edit.) Paris.

LEACH, E., 1961. *Re-Thinking Anthropology.* London.

LEVI-STRAUSS, C., 1958. *Anthropologie Structurale, Magie et Religion* in chapter X, 'L'efficacité Symbolique', originally published under same title in *Revue de l'Histoire des Religions,* 135, No. 1, 1949, pp. 5-27.

LEVY-BRUHL, L., 1922. *La Mentalité Primitive.* Paris.

1936. *Primitives and the Supernatural* (Trans., Clare). London.

LEWIS, I. M., 1963. 'Dualism in Somali Notions of Power'. *Journal of the Royal Anthropological Institute,* 93, 1. pp. 109-116.

LIENHARDT, R. G., 1961. *Divinity and Experience.* Oxford.

MACHT, D. I., 1953. 'An Experimental Pharmacological Appreciation of *Leviticus* XI and *Deut.* XIV'. *Bull. Hist. Medicine*, Vol. 27, pp. 444 ff.

MAIMONIDES, MOSES, 1881. *Guide for the Perplexed*. Translated by M. Friedlander, first edition. London.

MARSHALL, L., 1957. N/OW, *Africa*, 27, 3.

MARSHALL-THOMAS, E., 1959. *The Harmless People*. New York.

MARWICK, M. G., 1952. 'The Social Context of Cewa Witch Beliefs', *Africa*, 22, 3. pp. 215-33.

MAUSS, M., 1902-3. 'Esquisse d'une Théorie Générale de la Magie,' *L'Année Sociologique*, 1902-3, in collaboration with H. Hubert. Reprinted 1950 in *Sociologie et Anthropologie*. Paris.

MCNEILL & GAMER, 1938. *Medieval Handbooks of Penance*. New York.

MEAD, M., 1940. 'The Mountain Arapesh'. *Anthropological Papers*, American Museum of Natural History, Vol. 37.

MEEK, C. K., 1937. *Law and Authority in a Nigerian Tribe*. Oxford.

MEGGITT, M., 1962. *Desert People*. Sydney.

1964. 'Male-Female Relationships in the Highlands of Australian New Guinea. *American Anthropologist*, 2. 66. 4. pp. 204-23.

MICKLEM, NATHANIEL, 1953. *The Interpreter's Bible, II, Leviticus.*

MIDDLETON, J., 1960. *Lugbara Religion.*

MILNER, MARION, 1955. 'Role of Illusion in Symbol Formation'. *New Directions in Psychoanalysis*. (Edit. Klein, M.)

MORTON-WILLIAMS, P., 1960. 'The Yoruba Ogboni Cult in Oyo'. *Africa*, 30, 4.

MOULINIER, LOUIS, 1952. *Le Pur et l'Impur dans la Pensée des Grecs, d'Homère à Aristote*. Etudes et Commentaires, XI. Paris.

NADEL, S. F., 1957. 'Malinowski on Magic and Religion', in *Man and Culture*, Edit. R. Firth. London.

NAIPAUL, V. S., 1964. *An Area of Darkness*. London.

ONIANS, R. B., 1951. *Origins of European Thought about the Body, the Mind, etc.* Cambridge.

OSTERLEY & BOX. *The Religion of the Synagogue.*

PFEIFFER, R. H., 1957. *Books of the Old Testament.*

POLE, DAVID, 1961. *Conditions of a Rational Enquiry into Ethics.*

POSINKY, 1956. *Psychiatric Quarterly* XXX, p. 598.

POSPISIL, LEOPOLD, 1963. *Kapauku Papuan Economy*. New Haven.

RADCLIFFE-BROWN, R., 1933. *The Andaman Islanders*. Cambridge.

1939. *Taboo*, Frazer lecture.

RADIN, PAUL, 1956. *The Trickster, A Study in American Indian Mythology*. London.

1927. *Primitive Man as Philosopher*. New York.

RAUM, O., 1940. *Chagga Childhood*.

READ, H., 1955. *Icon and Idea, The Function of Art in the Development of Human Consciousness.*

READ, K. E., 1954. 'Cultures of the Central Highlands'. *South Western Journal of Anthropology*, 10. 1-43.

RICHARDS, A. I., 1956. *Chisungu.*

—— 1940. Bemba Marriage and Present Economic Conditions. *Rhodes-Livingstone Paper, No. 4.*

RICHTER, MELVIN, 1964. *The Politics of Conscience, T. H. Green and His Age.*

RICŒUR, P., 1960. *Finitude et Culpabilité.* Paris.

ROBINS, R. H., 1958. *The Yurok Language.*

ROBERTSON SMITH, W., 1889. *The Religion of the Semites.*

ROHEIM, G., 1925. *Australian Totemism.*

ROSE, H. J., 1926. *Primitive Culture in Italy.*

—— 1954. *Journal of Hellenic Studies*, 74, review of Moulinier.

SALIM, S. M., 1962. *Marshdwellers of the Euphrates Delta.* London.

SARTRE, J.-P., 1943. *L'Etre et le Néant.* 3rd edit.

—— 1948. *Portrait of an Anti-Semite.*

SAYDON, P. P., 1953. *Catholic Commentary on the Holy Scripture.*

SRINIVAS, M. N., 1952. *Religion & Society among the Coorgs of South India.* Oxford.

STANNER, W. E. H., *Religion, Totemism and Symbolism.*

STEIN, S., 1957. 'The Dietary Laws in Rabbinic & Patristic Literature'. *Studia Patristica*, Vol. 64, pp. 141 ff.

STEINER, F., 1956. *Taboo.*

TALCOTT-PARSONS, 1960. Chapter in *Emile Durkheim, 1858-1917. A Collection of Essays with Translations and a Bibiliography.* Edit. Kurt H. Wolff. Ohio.

TEMPELS, PLACIDE, 1952. *Bantu Philosophy.*

TURNBULL, C., 1961. *The Forest People.*

TURNER, V. W., 1957. *Schism and Continuity in an African Society.* Manchester.

—— 1962. *Chihamba, The White Spirit*, Rhodes-Livingstone Paper No. 33.

—— 1964. 'An Ndembu Doctor in Practice', chapter in *Magic, Faith and Healing* (Edit. Arikiev). Glencoe.

TYLOR, H. B., 1873. *Primitive Culture.*

VAN GENNEP, 1909. *Les Rites de Passage.* (English Translation 1960).

VANSINA, J., 1955. 'Initiation Rituals of the Bushong'. *Africa*, 25, 2, pp. 138-152.

—— 1964. 'Le Royaume Kuba'. Musée Royale de l'Afrique Centrale, *Annales-Sciences Humaines*, No. 49.

VAN WING, J., 1959. *Études Bakongo*, orig. pub. 1921 (Vol. I); 1938 (Vol. II). Brussels.

Bibliography

WANGERMAN, E., 1963. *Women in the Church. Life of the Spirit*, 27, 201, 1963.

WATSON, W., 1958. *Tribal Cohesion in a Money Economy*. Manchester.

WEBSTER, HUTTON, 1908. *Primitive Secret Societies. A Study in Early Politics and Religion*. 2nd edit. 1932. New York.

1948. *Magic, A Sociological Study*.

WESLEY, JOHN, 1826-7. *Works*, Vol. 5. 1st American Edition.

WESTERMARCK, EDWARD, 1926. *Ritual and Belief in Morocco*.

WHATELEY R., 1855. *On the Origin of Civilisation*.

WHATMOUGH, JOSHUA, 1955. *Erasmus*, 8, 1, pp. 618-9.

WILSON, BRIAN R., 1961. *Sects and Society*. London.

WILSON, MONICA, 1957. *Rituals and Kinship among the Nyakyusa*.

YALMAN, N., 1963. 'The Purity of Women in Ceylon and Southern India.' *Journal of the Royal Anthropological Institute*.

ZAEHNER, R. H., 1963. *The Dawn and Twilight of Zoroastrianism*.

Index

DICTIONARY OF ANALYTICAL PSYCHOLOGY

'Jung is a thinker who has affected us so deeply that many of his ideas have passed into common currency, and few people remember who was their originator.' – Philip Toynbee, *Observer*

A general description of psychological types, this dictionary sums up Jung's ideas and provides a valuable introduction for anyone who wants to understand Jung's typology and his ideas about the human personality. Jung believed that it is one's psychological type which from the outset determines and limits a person's judgment, so that his system of types is of central importance in his work.

Based on Jung's years of practical experience as a doctor and practising psychotherapist, the dictionary is a thoroughly practical work, rich in insights into the relation of the individual to the world, to people and to things.

C. G. JUNG

C. G. Jung (1875-1961), the Swiss psychiatrist and founder of Analytical Psychology, was an original thinker who made an immense contribution to the understanding of the human mind. In his early years he was a lecturer in psychiatry at the University of Zürich and collaborated with Sigmund Freud. He gave up teaching to devote himself to his private practice in psychiatry and to research. He travelled widely and was a prolific author, often writing on subjects other than analytical psychology, such as mythology, alchemy, flying saucers and the problem of time. Jung was also responsible for defining such influential and widely-used terms as the Collective Unconscious, Extraversion/Introversion and Archetypes.

ARK PAPERBACKS is an imprint of Routledge & Kegan Paul and can be ordered through your usual bookshop.

ARK PAPERBACKS

EROS AND CIVILIZATION
A PHILOSOPHICAL INQUIRY INTO FREUD

In this stimulating introduction to the philosophical aspects of Freud's ideas, Herbert Marcuse takes as his starting-point Freud's statement that civilization is based on the permanent subjugation of the human instincts. The methodical sacrifice of libido, and its rigid deflection to socially useful activities and expressions, is the basis of western civilization – especially its negative aspects and its repressive tendencies.

But, as Marcuse shows, Freud's own theory provides reasons for rejecting his identification of civilization with repression. Marcuse therefore applies the insights of Freud's speculative later work – his 'metapsychology': his theory of the instincts, his reconstruction of the prehistory of mankind – to an interpretation of the basic trends of civilization, stressing the philosophical and sociological implications of Freudian concepts.

HERBERT MARCUSE

Herbert Marcuse (1898-1979) was born and educated in Berlin. In 1934 he left Nazi Germany, and took refuge in the USA, where he taught at Columbia University. He then held appointments at Harvard, Brandeis and the University of California at San Diego, becoming known in the 1960s as the official idealogue of 'campus revolutions' in the USA and Europe. His books include *Reason and Revolution* (1941), *Soviet Marxism* (1958) and *One-Dimensional Man* (1964).

ARK PAPERBACKS is an imprint of Routledge & Kegan Paul and can be ordered through your usual bookshop.

ARK PAPERBACKS

GRAVITY AND GRACE

Simone Weil is widely recognized as one of the brilliant and original minds of twentieth-century France. Now known primarily as a religious thinker, she also devoted enormous energy in her formative years to work as a political activist and as a philosopher/teacher. Jewish by upbringing, in the later years of her life Simone Weil turned more and more towards Catholicism, having had several mystical experiences which deeply affected her thought.

Gravity and Grace shows her religious thoughts and ideas, drawn from many sources – Christian, Jewish, Indian, Greek and Hindu – and focusing on suffering and redemption. It brings the reader face to face with the profoundest levels of existence as Weil explores the relationship of the human condition to the realm of the transcendent.

SIMONE WEIL

Born in Paris in 1909, Simone Weil was educated at the Lycée Henri IV and the Ecole Normale Supérieure, where she was one of the first women students. She became a teacher, but also worked on farms and at the Renault car factory, choosing hard manual labour in order to experience the life of the working class. In 1936 she joined the Republican forces in the Spanish Civil War, returning from the front when she was badly burned in an accident. In 1942 she left France and joined the Free French in London. She died of tuberculosis the following year, refusing to eat more than the rations of those suffering Nazi occupation in her native France.

Her major works, published posthumously, include *Gravity and Grace*, *Waiting on God*, *The Need for Roots* and her *Notebooks*.

ARK PAPERBACKS is an imprint of Routledge & Kegan Paul and can be ordered through your usual bookshop.

ARK PAPERBACKS

THE LIBERAL AWAKENING 1815-1830

Elie Halévy's *A History of the English People*, the first volume of which appeared in Paris in 1912 and 1913, is regarded as a classic by historians of the nineteenth century. Published in six volumes, the *History* is now being reissued in paperback, and the first volume – *A History of the English People in 1815* – is already available.

This, the second volume, deals with the years from 1815 to 1830, covering the period of post-war recovery, the premierships of Castlereagh, Canning and Wellington, and taking us up to Catholic Emancipation.

Asa Briggs writes in the Foreword to Volume I: 'This new edition deserves to be widely read, for now, just as much as when it first appeared, the *History* raises fundamental questions about "England", a term which for Halévy included Wales, Scotland and Ireland, and about "history" as a discipline. ...History becomes not only a living subject, with the perspectives always changing, but an analytical study which tests the critical intelligence as much as it stimulates the imagination.'

ELIE HALEVY

Born in 1870, Elie Halévy studied at the Ecole Normale Supérieure, where he became the close friend of the philospher Alain. He first visited England in 1892, staying in London, Oxford and Cambridge, and going on to Ireland. During this visit, the first of what were to become annual events, he met Henry James, heard General Booth, leader of the Salvation Army, and went to the House of Commons, where he heard speeches by Gladstone, Balfour and Joseph Chamberlain.

For almost 40 years, from 1898 onwards, Halévy gave a course of lectures on British political ideas at the Ecole des Sciences Politiques in Paris. Apart from the classic *History of the English People*, Halévy's books include *The Growth of Philosophic Radicalism* and *Era of Tyrannies*, a collection of essays. He died in 1937.

ARK PAPERBACKS is an imprint of Routledge & Kegan Paul and can be ordered through your usual bookshop.

ARK PAPERBACKS

LES LIAISONS DANGEREUSES

'Even today *Les Liaisons* remain the one French novel that gives us an impression of danger: it seems to require a label on its cover reserving it for external use only.' – Jean Giraudoux

A great sensation at the time of first publication, with readers attempting to identify the real-life originals of the novel's characters, *Les Liaisons Dangereuses* reads as much the most 'modern' of eighteenth-century novels. It shows two experienced 'libertines', planning together to seduce two innocent characters whose lives they thus destroy. A married woman is also seduced and then abandoned as part of the game.

Brilliantly observed and vividly rendered in the letters which make up the novel, the characters take on a life of their own. Laclos lays bare layer after layer of their souls until we know them intimately. The novel is, as Richard Aldington writes in the Introduction, 'a tragic story well told, with a subtlety of psychological analysis not unworthy of a countryman of Stendhal, Flaubert and Balzac.'

PIERRE AMBROISE FRANCOIS CHODERLOS DE LACLOS

Born in Amiens in 1741, Pierre Ambroise Françoise Choderlos de Laclos entered the army at the age of eighteen and spent the next twenty years in various garrison towns without ever seeing battle. In 1779 he was sent to the island of Aix to assist in building a fort, and there wrote Les Liaisons Dangereuses. In 1786 he married Marie-Soulange Duperré and became an exemplary husband and father. He left the army in 1788, entered politics and was twice imprisoned during the Reign of Terror. He returned to the army as a general under Napoleon in 1800, and died at his post in Taranto, Italy, in 1803.

Laclos also wrote a treatise on the education of women and another on Vauban, but it is for *Les Liaisons Dangereuses*, his single masterpiece, that he is remembered.

ARK PAPERBACKS is an imprint of Routledge & Kegan Paul and can be ordered through your usual bookshop.